Ac. 1390, Conjugate Leaves 2+5, Extant pages 4 and 9

p. 4: The Conclusion of the Mathematical School Exercises in Greek and the Beginning of the Subachmimic Text, here John 10:7b-18

P. 9 John 11:43-55

Photograph by Pieterse Davison International Ltd.

CHESTER BEATTY MONOGRAPHS

No. 13

THE CHESTER BEATTY CODEX AC. 1390 MATHEMATICAL SCHOOL EXERCISES IN GREEK AND JOHN 10:7-13:38 IN SUBACHMIMIC

EDITED BY

WILLIAM BRASHEAR

WOLF-PETER FUNK

JAMES M. ROBINSON

RICHARD SMITH

PEETERS

LEUVEN - PARIS

1990

ISBN 90-6831-340-1
D. 1991/0602/44

CONTENTS

ACKNOWLEDGMENTS

I wish to express appreciation for the generous assistance during the preparation of this edition on the part of the personnel of the Chester Beatty Library, especially Wilfrid Lockwood, M.A., Director and Librarian, Cecily Golden, Personal Assistant to the Librarian and Secretary to the Library, and Dr. Patricia Donlon, at the time Reference Librarian and Curator of the Western Collection, our supervisor while working in the Garden Room. We are grateful not only for ready access to the manuscript itself, but also to the Registry of Accessions and the Archives, from which the acquisition details and thus the relation of this manuscript to the discovery of the Dishna Papers in 1952 could be worked out. We are also indebted to M.L. Bierbrier, Assistant Keeper in the Department of Egyptian Antiquities of the British Museum, for access to photographs made during the various stages of the conservation.

I also wish to express appreciation to the specialists enlisted in this edition: William Brashear for editing the mathematical exercises in Greek, Wolf-Peter Funk for writing the introduction, notes and critical apparatus for John 10:7-13:38 in Subachmimic, Richard Smith for preparing the indices, and Funk and Smith for together producing the transcription.

James M. ROBINSON

INTRODUCTION

AC. 1390

JAMES M. ROBINSON

INTRODUCTION

1. The Pachomian Library

The first Christian monastic order was founded in Upper Egypt by Pachomius early in the Fourth Century. What was left of its library was buried in the Seventh Century, to judge by the date of the latest material produced (a small papyrus roll with Horsiesios' Letter 3 in Sahidic). It was discovered late in 1952 in Upper Egypt near Dishnā, and hence is referred to locally as the Dishna Papers, though it has been known to scholars up to the present primarily as Bodmer Papyri.

The term Bodmer Papyri derives from the series of *editiones principes* entitled *Papyrus Bodmer I* (1954) through *Papyrus Bodmer XXIX* (1984), texts which come from the same manuscript discovery (except for P. Bodmer XVII) and are (with the exception of P. Bodmer VIII, a part each of P. Bodmer XXI and XXII, and a few scattered fragments) at the Bibliothèque Bodmer at Cologny near Geneva. By extension the term Bodmer Papyri has come to be used to refer to the manuscripts of the discovery, thus including materials from that discovery that happen to be at other repositories. This nomenclature has obscured the fact that a large part of the material is actually in the Chester Beatty Library (not a speak of minor parts at some six other locations).

Ḥasan Muḥammad al-Sammān and Muḥammad Khalīl al ʿAzzūzī are peasants from the hamlet Abū Manāʿ "Baḥrī", 5 km northeast of Fāw "Qiblī", the site of the headquarters of the Pachomian Monastic Order (variously spelt in the sources, e.g. Pabau [Greek], Pbaw or Pbow [Sahidic], or Phbow [Bohairic]), and 5.5 km northwest of Dishnā. Late in 1952 they were digging for *sabakh*, the fertilizer peasants like to collect especially from archaeological sites, some 300 meters out from the foot of the nearby Jabal Abū Manāʿ at al-Qurnah ("the corner"), when Ḥasan uncovered a large earthen jar containing the books. His brother-in-law ʿAbd al-ʿĀl ʿUmar al-Abbādī, in whose home he lived, trafficked in the books, selling most to a gold merchant of Dishnā, Riyāḍ Jirjis Fām, who over a period of time sold most of what he had to Phocion J. Tano, an established antiquities dealer of Cairo of Cypriot origin, who in turn sold the bulk of the discovery, estimated by Riyāḍ at some 33 books plus 10 small rolls the size of one's finger, to Martin Bodmer.

A sizable amount of the material held by Tano was acquired by Sir Chester Beatty: Ac. 2555 = a fragment from P. Bodmer II, 139/140 (John 19:25-28/30-32), a fragment from P. Bodmer XX (pp. 135,13-16/136,14-17), and an unidentified fragment; ac. 1389 = two-thirds of P. Bodmer XXI; and ac. 1390 published here. Probably of the same provenience are also ac. 1493 = *Apocalypse of Elijah*; ac. 1499 = a Greek grammar and a Graeco-Latin lexicon on Romans, 2 Corinthians, Galatians, and Ephesians; ac. 1501 = 4 sheets of one Psalms codex and 1 leaf of another Psalms text, perhaps an amulet; and perhaps ac. 2554 = P. Beatty Panopolitanus = an unbound, largely uninscribed codex with a few tax receipts of 339-47 A.D. from Panopolis (Achmim), constructed secondarily from two rolls containing the correspondence of the Strategus of the Panopolite nome from 298-300 A.D.

Minor segments, some passing through the hands of other middlemen and dealers, reached other repositories. The University of Mississippi, Oxford, Mississippi, acquired from a different dealer, Maguid Sameda, the Crosby-Schøyen Codex, of which 40 fragments were also acquired by the Bibliothèque Bodmer; Mississippi Coptic Codex II (half of P. Bodmer XXII); P. Rob. inv. 38 = a fragment of leaf 55/56 of P. Bodmer XXVI (Menander's *Aspis*, lines 487-98/524-32); and P. Duke inv. L 1 (ex P. Rob. inv. 201) = a fragment of Cicero *in Catilinam* from Codex Miscellani = P. Barcinonenses inv. 149-61 of which fragments were also located at the Bibliothèque Bodmer; and perhaps fragments joining those acquired by the University of Cologne as listed below.

The Institut für Altertumskunde of the University of Cologne acquired from Tano inv. 904 = P. Köln 3 = a fragment of P. Bodmer XXVI that joins the fragment which came to the University of Mississippi; and perhaps as many as four fragmentary texts that Cologne acquired in the same lot as that which included the fragment of P. Bodmer XXVI, for they join fragments that the University of Mississippi had acquired from the dealer from whom it acquired the material shared with the Bibliothèque Bodmer: P. Colon. inv. 906 that joins P. Rob. inv. 32 (scholia to *Odyssey* 1), P. Colon. inv. 901 that joins P. Rob. inv. 35 (Achilleus Tatios), P. Colon. inv. 902 = P. Köln 40 that joins P. Rob. inv. 43 (*Odyssey* 3-4) and P. Colon. inv. 903 that joins P. Rob. inv. 37 (a piece of ethnography or a philosophical treatise).

The Fundacio "Sant Lluc Evangelista" of Barcelona acquired Codex Miscellani = P. Barcinonenses inv. 149-61 = Cicero, *in Catilinam*, Psalmus Responsorius, Greek liturgical texts, and *Alcestis*. Fragments of Cicero, *in Catilinam*, were acquired by the Bibliothèque Bodmer as well as by the University of Mississippi.

The Palau Ribes Collection of the Seminario de Papirologia of the Facultad Teologica de San Cugat des Valles of Barcelona acquired P. Palau Ribes 181-83, administered by José O'Callaghan, S.J., who visited Dishnā in 1964-65 in search of ancient manuscripts. Hence this manuscript may stem from the same discovery.

There are probably further materials from the same discovery at some of these repositories that have not thus far been identified as such. The fragmentary book of Jannes and Jambres at the Chester Beatty Library is thought by its editor Albert Pietersma to be a likely candidate. But for lack of supporting evidence such items are not here included. And some of those listed only with hesitation may in fact not be from the same provenience. Furthermore, some of the material from the discovery may have reached still other repositories, not previously identified with the discovery, such as P. Yale Inv. 1779 = Psalms 76-77 in Coptic, a small roll in *rotuli* form (i.e. to be read with the roll held open vertically), acquired by Yale University, New Haven, Connecticut around 1962 from H.P. Kraus, the leading rare book dealer of New York. Yet lack of information about its provenience would make its inclusion here premature.

Some of this material found its way into other repositories: P. Bodmer VIII was presented to Pope Paul VI and is at the Vatican Library. The two codices of the University of Mississippi were acquired in 1981 by H.P. Kraus. The Crosby-Schøyen Codex was then acquired in 1981 by the Pax ex Innovatione Foundation of Houston. It was on deposit for reconservation, display and publication at the Institute for Antiquity and Christianity, Claremont, California, 1985-88, when it was returned to the Pax Foundation at Houston. It was sold at auction by Sotheby's in London on 6 December 1988 for £200,000 (some $322,000) to the Norwegian bibliophile Martin Schøyen, who added it to The Schøyen Collection, as ms. 193, London/Oslo. Mississippi Coptic Codex II is in a private collection in England. The Mississippi fragment of P. Bodmer XXVI (P. Rob. inv. 38) and the Mississippi fragment of Codex Miscellani = P. Barcinonenses inv. 149-61 now belong to the Library of Duke University in Durham, North Carolina (the latter P. Duke inv. L 1 [ex P. Rob. inv. 201]). Fragments of P. Barcinonenses inventory 149-61 that were located at the Bibliothèque Bodmer were given to the Fundacio "Sant Lluc Evangelista", which gave a fragment of P. Bodmer XXV it held to the Bibliothèque Bodmer. The Bibliothèque Bodmer also located and passed on to William H. Willis 40 fragments from the Crosby-Schøyen Codex. The fifteen that have been placed are from Melito of Sardis, *On the Passover*. All the fragments have recently been acquired from Duke University by Martin Schøyen and are currently in London being restored to their original locations in the codex.

The contents of the discovery, including the quite fragmentary items and those listed with hesitation, are as follows (they are papyrus codices, and in Greek, unless otherwise indicated):

1. Homer, *Iliad*, Book 5 = P. Bodmer I, a roll on the verso of a roll of documentary papyri, = P. Bodmer L.

2. Homer, *Iliad*, Book 6 = P. Bodmer I, a roll on the verso of the same roll of documentary papyri, = P. Bodmer L.

6. Gospel of John = P. Bodmer II + a fragment from the Chester Beatty Library, ac. 2555, + P. Köln 214, = P. 66.

4. Gospel of John and Genesis 1:1-4:2 in Bohairic = P. Bodmer III.

5. Menander, *Samia*, *Dyskolos*, *Aspis* = P. Bodmer XXV, IV, XXVI + P. Barc. 45 + Cologne inv. 904 = P. Köln 3 + P. Rob. 38.

6. *Nativity of Mary* = *Apocalypse of James* (*Protevangelium of James*); Apocryphal Correspondence of Paul with the Corinthians; *Odes of Solomon* 11; the Epistle of Jude; Melito of Sardis *On the Passover*; a fragment of a liturgical hymn; the *Apology of Phileas*; Psalms 33-34; 1 and 2 Peter = P. Bodmer V; X; XI; VII; XIII; XII; XX (+ a fragment from the Chester Beatty Library, ac. 2555); IX, VIII.

7. Proverbs in Palaeo-Theban on parchment = P. Bodmer VI.

8. Gospels of Luke and John = P. Bodmer XIV-XV = P 75.

9. Exodus 1:1-15:21 in Sahidic on parchment = P. Bodmer XVI. (P. Bodmer XVII is generally agreed not to come from the same discovery).

10. Deuteronomy 1:1-10:7 in Sahidic = P. Bodmer XVIII.

11. Matthew 14:28-28:20 + Romans 1:1-2:3, both in Sahidic on parchment, = P. Bodmer XIX.

12 .Joshua in Sahidic = P. Bodmer XXI + Chester Beatty ac. 1389.

13. Jeremiah 40:3-52:34; Lamentations; Epistle of Jeremy; Baruch 1:1-5:5, all in Sahidic on parchment, = P. Bodmer XXII + Mississippi Coptic Codex II.

14. Isaiah 47:1-66:24 in Sahidic = P. Bodmer XXIII.

15. Psalms 17-118 = P. Bodmer XXIV.

16. Thucydides; Suzanna; Daniel; Moral Exhortations = P. Bodmer XXVII, XLV, XLVI, XLVII.

17. A satyr play on the confrontation of Heracles and Atlas, a papyrus roll, = P. Bodmer XXVIII.

18. Codex Visionum = P. Bodmer XXIX-XXXVIII. (For P. Bodmer XXXIX see the inventory of Pachomian material below).

19. Song of Songs in Sahidic on parchment = P. Bodmer XL.

20. *The Acts of Paul*, Ephesus Episode, in Subachmimic, = P. Bodmer XLI.

21. 2 Corinthians in Coptic (dialect and material unknown) = P. Bodmer XLII.

22. An Apocryphon in Coptic (identity, dialect and material unknown) = P. Bodmer XLIII.

23. Daniel in Bohairic = P. Bodmer XLIV.

24. Fragments of the *Iliad* from a papyrus roll = P. Bodmer XLVIII.

25. Fragments of the *Odyssey* from a papyrus roll = P. Bodmer XLIX.

26. Mathematical exercises in Greek; John 10:7-13:38 in Sabachmimic = Chester Beatty ac. 1390.

27. *The Apocalypse of Elijah* in Sahidic = Chester Beatty ac. 1493 = P. Chester Beatty 2018.

28. A Greek grammar; a Graeco-Latin lexicon on Romans, 2 Corinthians, Galatians, Ephesus = Chester Beatty ac. 1499.

29. Psalms 72:6-23, 25-76:1; 77:1-18, 20-81:7; 82:2-84:14; 85:2-88:20 = Chester Beatty ac. 1501 = P. Chester Beatty XIII = Rahlfs 2149.

30. Psalms 31:8-11; 26:1-6, 8-14; 2:1-8 = Chester Beatty ac. 1501 = P. Chester Beatty XIV = Rahlfs 2150.

31. Tax receipts of 339-47 A.D. from Panopolis (Achmim) in a largely uninscribed and unbound quire contructed from two papyrus rolls with correspondence of the Strategus of the Panopolitan nome of 298-300 A.D. = P. Beatty Panopolitanus = Chester Beatty ac. 2554.

32. Melito of Sardis *On the Passover*; 2 Maccabees 5:27-7:41; 1 Peter; Jonah; a homily or hymn = The Crosby-Schøyen Codex.

33. Scholia to the *Odyssey* 1 from a papyrus roll = P. Rob. inv. 32 + P. Colon. inv. 906.

34. Achilleus Tatios from a papyrus roll = P. Rob. inv. 35 + P. Colon. inv. 901.

35. *Odyssey* 3-4 from a papyrus roll = P. Rob. inv. 43 + P. Colon. inv. 902.

36. A piece of ethnography or a philosophical treatise from a papyrus roll = P. Rob. inv. 37 + P. Colon. inv. 903.

37. Cicero, *in Catilinam*; Psalmus Responsorius; Greek liturgical text; *Alcestis*, all in Latin except the Greek liturgical text, = Codex Miscellani = P. Barcinonenses inv. 149-61 + P. Duke inv. L 1 [ex P. Rob. inv. 201].

38. Gospels of Luke; John; Mark, all in Sahidic = P. Palau Ribes 181-183.

The total quantity of material would involve what remains of some 38 books. They consist of 9 Greek classical papyrus rolls (numbers 1, 2, 17, 24, 25, 33-36) and 29 codices (numbers 3-16, 18-23, 26-32, 37, 38). The codices may be subdivided as follows: 22 are on papyrus (numbers 3-6, 8, 10, 12, 14-16, 18, 20, 23, 26-32, 37, 38), 5 on parchment (numbers 7, 9, 11, 13, 19), and of 2 the Bibliothèque Bodmer has not divulged the material (numbers 21, 22). 10 are in Greek (numbers 3, 5, 6, 8, 15, 16, 18, 29-31), 2 in Greek and Latin (numbers 28, 37), and 1 in Greek and Subachmimic (number 26). 16 are in Coptic (numbers 4, 7, 9-14, 19-23, 27, 32, 38), of which 10 are in Sahidic (numbers 9-14, 19, 27, 32, 38), 2 in Bohairic (numbers 4, 23), 1 in Palaeo-Theban (number 7), 1 in Subachmimic (number 20), and of 2 the Bibliothèque Bodmer has not divulged the dialect (numbers 21, 22). 2 are non-Christian (numbers 5, 31), 23 Christian (numbers 3, 4, 6-15, 18-23, 27, 29-30, 32, 38) and 4 partly each (numbers 16, 26, 28, 37). 12 contain something from the Old Testament (numbers 7, 9, 10, 12-16, 19, 23, 29, 30), 6 something from the New Testament (numbers 3, 8, 11, 21, 26, 38) and 3 something from each (numbers 4, 6, 32).

A distinctive part of this discovery consists of archival copies of official letters of Abbots of the Pachomian Monastic Order:

1. Pachomius' Letter 11b in Sahidic, a small parchment roll, = P. Bodmer XXXIX.
2. Pachomius' Letters 9a, 9b, 10 11b, from a papyrus codex, in Sahidic, = Chester Beatty Glass Container No. 54 = ac. 2556.
3. Pachomius' Letters 1-3, 7, 10, 11a in Greek, a small parchment roll in *rotuli* format, = Chester Beatty Ms. W. 145 + Cologne inv. 3288 = P. Köln 174 = three fragments from Letter 7.
4. Theodore's Letter 2 in Sahidic, a small parchment roll in *rotuli* format, = Chester Beatty Library ac. 1486.
5. A second copy of Theodore's Letter 2, a small parchment roll in *rotuli* format in an unidentified private German collection, published by Martin Krause.
6. Horsiesios' Letter 3 in Sahidic, a small papyrus roll, = Chester Beatty Library ac. 1494.

7. Horsiesios' Letter 4 in Sahidic, a small papyrus roll, = Chester Beatty Library ac. 1495.

8. Pachomius' Letter 8 in Sahidic, a small parchment roll, = Cologne inv. 3286 = P. Colon. Copt. 2 = P. Köln ägypt. 8.

9. Pachomius' Letters 10-11a in Sahidic, a small parchment roll, = Cologne inv. 3287 = P. Colon. Copt. 1 = P. Köln ägypt. 9.

All these letters, previously completely unknown or known only in Jerome's Latin translation of 404 A.D., have suddenly emerged at the same time and (with one exception, item 5) at the same repositories as did the Dishna Papers. Most of them are apparently what Riyāḍ referred to as the small rolls, which Tano told him contained letters. In one instance (Cologne) the repository has a record ascribing its Pachomian material to the Nag Hammadi region, Tano's favorite vague but advantageous designation of the provenience of the the Bodmer Papyri. Two of the repositories (Cologne and Dublin) are known to have acquired at least some of their Pachomian material in the same acquisition with other Dishna Papers. The site of the discovery is in full view of the Pachomian Order's headquarters Monastery at Fāw Qiblī 5 km away. The site, well above the inundation level (in distinction from Fāw Qiblī itself), is in the remote area at the foot of the cliff to which funeral processions moved from the monastery at the death of a monk. These considerations make it probable that the Pachomian Archive and thus the discovery as a whole made up the Archives of the Pachomian Monastic Order.

The First Greek Life of Pachomius 63 gives some information about how books were viewed in the Pachomian Order[1]:

> He [Pachomius] also used to teach the brothers not to give heed to the splendor and the beauty of this world in things like good food, clothing, a cell, or a book outwardly pleasing to the eye.

The First Greek Life of Pachomius 59 gives some impression of a Pachomian Library[2]:

> No one would do anything in the house without permission from those in charge, not even visit a brother in his cell. In each house, the housemaster or the second keeps all the surplus clothings locked in a cell until the brothers need them to wash and put on again those they

[1] *Pachomian Koinonia* 1: *The Life of Saint Pachomius*, 2: *Pachomian Chronicles and Rules*, tr. Armand Veilleux (Cistercian Studies 45 and 46; Kalamazoo, MI: Cistercian Publications, 1980 and 1981), 1.341. See the thorough analysis by Clemens SCHOLTEN, "Die Nag-Hammadi-Texte als Buchbesitz der Pachomianer", *Jahrbuch für Antike und Christentum* 31 (1988) 144-172.

[2] *Pachomian Koinonia*, 1.338-339.

> are using. The books, which were in an alcove, were also under the care of these two. The brothers have no money, still less anything of gold; some of them died having never known such things. Only those entrusted with a ministry used money; and when they returned to the monastery they kept nothing with themselves for a single day and gave everything to the steward until they might go out again. And all that government is written in detail in the book of the stewards.

What is here referred to as the book of the stewards is apparently the extant *Precepts*[3], where a rather massive literacy program is envisaged and occasional references to books and to the Library occur (*Precepts* 139, 140, 82, 100, 101)[4]:

> Whoever enters the monastery uninstructed shall be taught first what he must observe; and when, so taught, he has consented to it all, they shall give him twenty psalms or two of the Apostle's epistles, or some other part of the Scripture. And if he is illiterate, he shall go at the first, third, and sixth hours to someone who can teach and has been appointed for him. He shall stand before him and learn very studiously with all gratitude. Then the fundamentals of a syllable, the verbs, and nouns shall be written for him, and even if he does not want to, he shall be compelled to read.
>
> There shall be no one whatever in the monastery who does not learn to read and does not memorize something of the Scriptures. [One should learn by heart] at least the New Testament and the Psalter.
>
> No one shall have in his own possession little tweezers for removing thorns he may have stepped on. Only the housemaster and the second shall have them, and they shall hang in the alcove in which books are placed.
>
> No one shall leave his book unfastened when he goes to the *synaxis* or to the refectory.
>
> Every day at evening, the second shall bring the books from the alcove and shut them in their case.

Ac. 1390 is itself a school-boy's Greek exercises in solid geometry that rendered the rest of the quire of little financial value, the kind of material a Pachomian monastery might be able to afford. On the empty pages a few chapters of the Gospel of John in Coptic were written in a non-literary, cursive hand, beginning in the middle of a sentence. This may be explainable as the place where the mutilated text being copied had begun. Or perhaps ac. 1390 was one in a series of cheap writing materials, the only one to have survived, onto which the complete Gospel was copied. The preceding (lost) writing surface on which the

[3] *Pachomian Koinonia*, 2.414-415.

[4] *Pachomian Koinonia*, 2.166.260-262.

Gospel was being copied would have ended in the middle of a verse, which would explain why the text that has survived begins there, just where the other happened to break off. Thus ac. 1390 may give some insight into the limitations of the monastic effort to build its collection.

A similar impression of primitiveness may be conveyed by other materials such as the largely uninscribed ac. 1499 containing a Greek grammar and a Graeco-Latin lexicon for deciphering Pauline epistles, as well as ac. 2554, a largely uninscribed and unbound folded stack of sheets constructed by pasting face to face two inscribed rolls and cutting them into the sheets of a quire on which administrative records had begun to be inscribed, with the result that the material would have but little commercial value. A different kind of improvisation would seem to be reflected in the secondary binding together of a number of tractates into a codex of sorts (number 6 in the inventory).

It may be no coincidence that much of the material of the highest quality in the collection is older than the Pachomian Order itself, suggesting that it entered the Library as gifts from outside, perhaps contributed by prosperous persons entering the Order. This might be the most obvious way to explain non-Christian texts in a monastic library, such as the Homeric and Menander material. But some such explanation is also needed for such excellent early Greek New Testament texts as P. Bodmer II (P. 66, the Gospel of John), and P. Bodmer XIV-XV (P 75, the Gospels of Luke and John), where one might even think of Athanasius living in hiding with the Order while in exile as the source of such gifts.

The bulk of Christian codices date from the first century of the Pachomian Order's existence, namely the early Fourth to the early Fifth Century, and some do present the competence of a trained scriptorium, though without adornment. And there is no specific indication that they came from a single scriptorium or that such a scriptorium belonged to the Order.

Later texts in the collection, such as some of the copies of letters from Pachomian Abbots, again suggest, in the primitiveness of the material employed (otherwise unusable leather scraps or the skin of an animal's leg), that the usual standards of a scriptorium were lacking.

The presence of relatively unskilled products alongside of relatively professional codices may indicate a plurality of places of origin, and perhaps a contrast between what was produced within the Order and what came from outside.

If discipline relaxed and the demand for reading material waned with the passage of time, as the center of learning shifted to the Monastery of Shenouda at Sohag, such a Pachomian collection could have become more a geniza than an active library. The identity of increasingly fragmentary items in the collection would be lost from sight, especially in the case of the old non-Coptic material, if one may assume that the Greek House did not retain its original vigor at Fāw Qiblī, once the Order had a monastery near Alexandria where Greek-speaking monks would be more at home. It would have been enough that the remains represented the venerated relics of the beginnings of the Order, worthy to be included along with copies of official letters of the early Abbots (about the only thing they continued to copy), in a jar no doubt intended to rescue for posterity the surviving symbols of continuity with the Order's legendary past.

This is illustrated in another way by the fate of the excellent third-century copy of the Gospels of Luke and John (P 75): It was rebound in late antiquity by pasting fragmentary leaves of the quire together as cartonnage to thicken the leather cover, and by sewing the binding thongs through the inner margin of what remained of the quire so near the writing that the codex could not be opened wide enough to be actually read. One is inclined to think that the codex had become a relic, the Library a Museum, or, in view of the copies of official Pachomian texts, an Archive.

Except for the copies of official Pachomian letters, datings as late as the Fifth Century are not strongly represented. For it is usually mentioned by editors in the spectrum of Fourth or Fifth Century[5]. In one of these cases, P. Bodmer VI, the editor has subsequently expressed a preference for the Fourth Century[6], and in the case of the second, P. Bodmer XIX, he has narrowed his dating to the end of the Fourth Century[7]. In the case of the third, P. Bodmer XXII = Mississippi Coptic Codex II, William H. Willis has preferred the Fourth Century[8].

[5] *Papyrus Bodmer VI. Livre des Proverbes*, ed. and tr. by Rodolphe KASSER (Corpus Scriptorum Christianorum Orientalium 194-95, Scriptores Coptici 27-28; Louvain: Secrétariat général du CorpusSCO, 1960), 2.5.

[6] Rodolphe KASSER, "Les dialectes coptes", *Bulletin de l'Institut français d'archéologie orientale* 73 (1973) 81: "The manuscript [P. Bodmer VI] that attests [the Palaeo-Theban dialect] is of the Fourth (perhaps the Fourth-Fifth) Century".

[7] *Papyrus Bodmer XIX, Évangile de Mattieu XIV, 28-XXVIII, 20; Épître aux Romains I, 1-II, 3 en sahidique*, ed. by Rodolphe KASSER (Cologny-Geneva: Bibliothèque Bodmer, 1962), p. 8, n. 3: "Assuming that the copy was effected at the latest at the end of the Fourth Century..."

[8] William H. WILLIS, "The New Collection of Papyri at the University of Mississippi", *Proceedings of the IX International Congress of Papyrology* (Oslo: Norwegian Universities

The Apocalypse of Elijah = Chester Beatty Library ac. 1493 has also been dated to the Fourth of Fifth Cnetury[9]. P. Palau Ribes 181-183 has been dated to the earth Fifth Century or later[10], but it is one of the codices least certain to belong to the collection. The Crosby-Schøyen Codex has been dated variously from the Third to the Sixth Century, but the most recent and perhaps most authoritative has been to the end of the Fourth Century[11]. Similarly contested has been A.F. Shore's Fourth Century dating of P. Bodmer XXI = Chester Beatty Library ac. 1389 on the part of Martin Krause, whose criteria for dating to the Fifth Century have however been largely put in question by Eric G. Turner's standard work of 1977, *The Typology of the Early Codex*[12].

Press, 1961), p. 391. Willis also wrote in a letter of 25 ii 80: "I know of nothing in the Bodmer-Mississippi-Beatty-Barcelona group that requires a date of deposit later than A.D. 400".

[9] Albert PIETERSMA and Susan Turner COMSTOCK, with Harold W. ATTRIDGE, *The Apocalypse of Elijah based on P. Chester Beatty 2018* (Texts and Translations 19, Pseudepigrapha Series 9; Chico: Scholars Press, 1981).

[10] Hans QUECKE, *Das Lukasevangelium saïdisch. Text der Handschrift PPalau Rib. Inv.-Nr. 181 mit den Varianten der Handschrift M 569* (Papyrologica Castroctaviana, Studia et textus 6, directed by José O'Callaghan, S.J.; Barcelona: Papyrologica Castroctaviana, 1977) p. 13 (see also p. 88):

> Admittedly I would put a question mark beside "first half", since the script of P stands in a certain proximity to that of the Vienna Dioskurides codex that was written around 512.

But in "Eine koptische Handschrift des 5. Jahrhunderts III (PPalau Rib. Inv.-Nr. 183)", *Studia Papyrologica* 20 (1981) 10 Quecke comments:

> Nonetheless in my quite personal opinion very good reasons can be given for the view that the manuscript is no doubt to be moved rather toward the beginning than to the end of the Fifth Century.

[11] William H. WILLIS, "The New Collections of Papyri at The University of Mississippi", p. 389, Third Century, and in a letter of 30 vii 84 as second half of the Second Century. Colin H. ROBERTS, cited by Willis in his letter, late Second or early Third Century. Allen CABANISS, "The University of Mississippi Coptic Papyrus Manuscript: A Paschal Lectionary?" *New Testament Studies* 8 (1961) 71: "*c.* A.D. 300". Eric G. TURNER, *The Typology of the Early Codex* (Philadelphia: University of Pennsylvania, 1977), p. 137, "Third or Fourth" Century. Stuart George HALL, *Melito of Sardis, On Pascha and Fragments* (Oxford Early Christian Texts; Oxford: Clarendon, 1979, xlv, "probably" the Sixth Century, and XVII, n. 8:

> But experts to whom I have shown my photocopy, including Professor T. Orlandi, suggest the Sixth, or at earliest the Fifth, Century.

Kurt and Barbara ALAND, The Text of the New Testamenty (Grand Rapids, MI: William B. Eerdmans, and Leiden: E. J. Brill, 1987), p. 197, "from the third century (probably *ca.* 400)".

[12] A.F. SHORE, *Joshua I-VI and Other Passages in Coptic, edited from a Fourth-Century Sahidic Codex in the Chester Beatty Library, Dublin* (Chester Beatty Monographs 9; Dublin: Hodges Figgis and Co. Ltd, 1963), p. 11. Similarly Rodolphe KASSER, *Papyrus Bodmer XXI, Josué VI, 16-25, VII, 6-XI, 23, XXII, 1-2, 19-XXIII, 7, 15-XXIV, 23 en*

Thus it is apparent that the Fifth Century is only weakly and uncertainly represented among the Dishna Papers. The source of supply seems to have been drying up, or new production was being attracted to the White Monastery.

When one turns to the official Pachomian materials, the archival copies of letters of Pachomian Abbots, the situation is the converse. Whereas the earliest material is by the nature of the case no earlier than the Fourth Century, only one text (number 3 in the listing above) is dated by its editor simply to the Fourth Century[13], and only one (number 1) to the Fourth or Fifth Century[14]. One (number 5) is dated simply to the Fifth Century[15], and two (numbers 8 and 9) to the Fifth or Sixth Century[16]. Three (numbers 2, 4 and 7) are dated simply to the Sixth Century[17], and one (number 6) simply to the Seventh Century[18].

sahidique (Cologny-Geneva: Bibliothèque Bodmer, 1963), p. 5: "... probably of the Fourth Century". Martin KRAUSE, "Schätze aus dem zweiten grossen Fund koptischer Handschriften", *Orientalische Literaturzeitung* 62 (1967) 442. Eric C. TURNER, *The Typology of the Early Codex* (Oxford Early Christian Texts; Oxford: Clarendon, 1977).

[13] Hans QUECKE, "Die Briefe Pachoms", *Zeitschrift der Deutschen Morgenländischen Gesellschaft*, Supp. 2 1974 (= 18. Deutscher Orientalistentag 1972), p. 98, n. 13, advocates the Fourth Century, and reports that the same date was already proposed by T.C. SKEAT in a letter of 17 xii 70. Hans QUECKE, *Die Briefe Pachoms: Griechischer Text der Handschrift W. 145 der Chester Beatty Library, eingeleitet und herausgegeben* (Textus Patristici et Liturgici 11; Regensburg: Friedrich Pustet, 1975).

[14] Tito ORLANDI, A. de VOGÜÉ, Hans QUECKE and James GOEHRING, *Pachomiana Coptica*, in the press. The dating is from an early draft of the typescript by de Vogüé.

[15] Martin KRAUSE, "Der Erlassbrief Theodors", *Studies Presented to Hans Jakob Polotsky*, ed. Dwight W. YOUNG (Beacon Hill, East Gloucester, MA: Pirtle and Polson, 1981), pp. 220-38 and Plate 6, especially p. 221: "with every reservation, the Fifth Century".

[16] Number 8: Dieter KURTH, Heinz-Jozef THISSEN and Manfred WEBER, *Kölner Ägyptische Papyri (P. Köln ägypt.)* 1 (Abhandlungen der Rheinisch-Westfälischen Akademie der Wissenschaften, Sonderreihe Papyrologica Coloniensia 9; Cologne and Opladen: Westdeutscher Verlag, 1980), pp. 100-02. Hans QUECKE, "Die Briefe Pachoms", p. 97, cited with apparent approval the dating "Fifth or Sixth Century" of Alfred Hermann in his very inadequate *editio princeps*, "Homilie in sahidischem Dialekt", *Demotische und Koptische Texte* (Papyrologica Coloniensia 2; Wissenschaftliche Abhandlungen der Arbeitsgemeinschaft für Forschung des Landes Nordshein-Westfalens; Cologne and Opladen, 1968), pp. 82-85 and Plate 3, especially p. 82. Number 9: KURTH, THISSEN and WEBER, *Kölner Ägyptische Papyri (P. Köln ägypt.)* 1, pp. 103-08. Hans QUECKE, "Die Briefe Pachoms", p. 97, cited with apparent approval the dating "Fifth or Sixth Century" as that of Angelicus KROPP, O.P., "Ein Märchen als Schreibübung", *Demotische und Koptische Texte*, pp. 69-81 and Plates 1-2, especially p. 81, where KROPP wrote "end of the Fifth Century".

[17] Number 2: Hans QUECKE, "Ein neues Fragment der Pachombriefe in koptischer Sprache", *Orientalia* 43 (1974) 66-72, especially p. 67, "probably from the Sixth Century". Number 4: Hans QUECKE, "Ein Brief von einem Nachfolger Pachoms (Chester Beatty Library Ms. Ac. 1486)", *Orientalia* 44 (1975) 426-33 and Plate 42, especially p. 427,

Thus it is clear that the letters of the Pachomian Abbots continued being copied much later than did the codices themselves, and represent the clearest indication of the interest of those responsible for the continuation of the Library or Archive in its latest period and hence for its ultimate burial[19].

Perhaps these relics were buried for safe keeping in connection with the imposition of Chalcedonian orthodoxy on the traditionally Monophysite Order, as the dating of the latest materials might suggest. A Coptic manuscript (M 579 of the Pierpont Morgan Library) dated to 822-23 A.D. does in fact report the abandonment of the monastery by the Monophysite monks in protest over the Chalcedonian takeover[20].

2. The Acquisition by the Chester Beatty Library

Sir Alfred Chester Beatty (1875-1968)[21] had his main residence in London at Baroda House in Kensington Palace Gardens until 1950, when he moved to Dublin. Actually he had customarily wintered at the "Blue House" he had bought in Giza near Cairo. But this habit came to an end with the move to Dublin in May 1950, after which he tended to winter at Nice in southern France. He acquired both a home in

"probably of the Sixth Century". Number 7: Tito Orlandi, "Nuovi Testi Copti Pacomiani", *Commandements du Seigneur et Libération évangélique* (Studia Anselmiana 70; Rome: Editrice Anselmiana, 1977), pp. 241-43, esp. p. 242, where he refers to Guglielmo Cavallo for a dating "a bit older than that of the preceding roll" (see the following note). Hans Quecke, "Eine Handvoll pachomianischer Texte", *Zeitschrift der Deutschen Morgenländischen Gesellschaft*, Supp. 3,1 1977 (= 19. Deutscher Orientalistentag 1975), pp. 221-29, especially p. 222, lists the Sixth Century.

[18] Orlandi, "Nuovi Testi Copti Pacomiani", p. 241, cites Guglielmo Cavallo for a dating to the Seventh Century. Quecke, "Eine Handvoll pachomianischer Texte", p. 222:

> The hand is a very artificial uncial, which one would like to place considerably later [than a Sixth Century dating, see the preceding note].

[19] Already Hans Quecke has recognized the non-accidental nature of the five Pachomian texts acquired by the Chester Beatty Library, "Eine Handvoll pachomianischer Texte", p. 221:

> It is to be suspected that the five pieces belong together, and thus, as it were, present a "hoard" of Pachomian material. The five Pachomian pieces can indeed hardly have come together accidentally in the Chester Beatty Library.

And Tito Orlandi, "Nuovi Testi Copti Pacomiani", p. 241, considers the material to come from "the library of a Pachomian monastery".

[20] K.H. Kuhn, *A Panegyric on Apollo Arichimandrite of the Monastery of Isaac by Stephen Bishop of Heracleopolis Magna* (Corpus Scriptorum Ecclesiasticorum Orientalium 394-95; Scriptores Coptic 39-40; Louvain: Secrétariat du CorpusSCO, 1978).

[21] A.J. Wilson, *The Life and Times of Sir Alfred Chester Beatty* (London: Cadogan Publications Ltd., 1985), presents an informed biography, including only a few brief treatments of the bibliophile dimension of Beatty's activity.

Dublin itself in Ailesbury Road and the property at 20 Shrewsbury Road for the Library, where a home for the first Librarian James Vere Stewart Wilkinson (1885-1957) was also built. The first public building was constructed in 1952-53. The opening ceremonies took place on 11 viii 54, a month after Beatty was knighted by the Queen. Then a second building at the rear was constructed in 1955-56 and opened as a Gallery for the public in 1957, making possible a display of previously undisplayed papyri[22]. In 1975 the Irish government enlarged it by adding a second storey and an extension on the front with a new glass facade. This conformed to the Chester Beatty Bequest, according to which the buildings and their contents were left to a Board of Trustees to be administered on behalf of the Irish people[23]. The property, marked "The Chester Beatty Library and Gallery of Oriental Art", has a reading room for those studying manuscripts called the "Garden Library", a large room at the back of the first building extending into the garden that separates it from the newer Gallery, to which manuscripts are brought temporarily for study[24].

The library of Sir Chester Beatty has been famous since the early 1930s for *The Chester Beatty Biblical Papyri*, a dozen early Greek

[22] T.C. SKEAT, "Papyri from Panopolis in the Collection of Sir Chester BEATTY", *Proceedings of the IX International Congress of Papyrology, Oslo, 19th-22nd August, 1958*, Association internationale de papyrologues, ed. L. Amundsen and V. Skandland ([Oslo]: Norwegian Universities, 1961), p. 194:

> In 1957 Sir Chester Beatty opened the new wing of his celebrated Library in Dublin, and this extension gave him a long-awaited opportunity for setting out and arranging various sections of his manuscript collections, including a number of papyri which had been acquired at various times during the past twenty or thirty years, and which could now for the first time be examined in detail.

See also the guide booklet by the Librarian R.J. HAYES, *The Chester Beatty Library, Dublin*, 1958, reprint 1971, p. 5:

> The collections were housed for many years in Baroda House, Kensington Palace Gardens, London, until in 1953, having moved his residence to Dublin, he built a special library building to hold the wonderful collections of Oriental and Western printed books and manuscripts which he had brought together during the previous forty years. In 1957 an additional large exhibition gallery was added to enable more of the treasures to be displayed. These buildings are situated in a pleasant surburban setting in Shrewsbury Road, Dublin.

[23] Brian KENNEDY, 'Sir Chester Beatty — Friend and Patron of Ireland", *The Irish Times*, 7 ii 85, p. 10. This article is based on Kennedy's 1985 M.A. Thesis at University College Dublin on Beatty's activity in Ireland.

[24] Tito ORLANDI, "Les Manuscrits coptes de Dublin, du British Museum et de Vienne", *Le Muséon* 89 (1976) 326.

papyri that were the sensational biblical discovery of that generation[25]. For Chester Beatty Biblical Papyri I (the Gospel and Acts), II (the Pauline Epistles) and III (the Book of Revelation), usually known as P 45, P 46 and P 47 respectively, attested the text of the New Testament back behind the great Fourth Century uncial manuscripts Sinaiticus and Vaticanus, which had seemed a century ago to be as far back as one could reasonably hope to have manuscript evidence. For P 45-47 went back to the early or middle Third Century, behind the Diocletianic persecutions, which included in the imperial decree *traditio codicum*, the "turning in" of Christian books for destruction. One had hardly dared to hope to recover biblical manuscripts older than this book-burning holocaust.

A generation later Bodmer Papyri II (the Gospel of John) and XIV-XV (the Gospels of Luke and John) were again dramatic Third Century manuscripts which tended to shift the focus of attention from Dublin to Geneva. But what has not been generally realized is that Sir Chester Beatty also acquired a number of Dishna Papers. It is Beatty's acquisition of texts from this discovery which is here to be traced. For accession number 1390, published in the present volume[26], is the clearect instance of a text at the Chester Beatty Library emanating from the discovery near Dishnā. Indeed in the Registry of Accessions at 1390 there is stapled to the left-hand page, usually left empty for supplemental notes, a typed note identifying the text as emanating from the Dishnā discovery, a note probably written by the antiquities dealer Tano himself, to judge by its wooden English style and orthography:

> Small Village DESHNA just after NAGHI HAMADI about 2 hours before LUXOR by train. Probably from a Library of a Monastery. Found in a jar in a cemetery.

Beatty had been acquiring papyri and other antiquities from Tano for a number of years. A letter from "Ph. John Tano" dated 9/2/43 bears on its letterhead "House Founded in 1879: Nicolas Tano". Tano gave as his address for correspondence: Phocion John Tanos [the Greek

[25] Frederic G. KENYON, *The Chester Beatty Biblical Papyri: Descriptions and Texts of Twelve Manuscripts on Papyrus of the Greek Bible* (London: Emery Walker, 1933-41), plus a few supplemental volumes. See most recently A. PIETERSMA, "New Greek Fragments of Biblical Manuscripts in the Chester Beatty Library", *Bulletin of the American Society of Papyrologists* 24:1-2 (1987) 37-61.

[26] Though unpublished, this manuscript has been mentioned several times: R.J. HAYES, *The Chester Beatty Library, Dublin*, p. 10; A.F. SHORE, *Joshua I—VI and Other Passages in Coptic*, p. 12, n. 2; Tito ORLANDI, "Les Manuscrits coptes de Dublin, du British Museum et de Vienne", p. 326.

spelling of the family name], care of Joannides and Acker, 35 Liperte Street, Nicosia, Cyprus". On 8 September 1947 the Chester Beatty Library paid Tano £ 24 for four leaves from a codex, to be paid care of the Ottoman Bank, Famagusta, Cyprus. On 16 April 1948 Tano sent Beatty four wooden tablets through the good offices of his brother-in-law, William Acker, an offiicer in the RAF. In 1950 Beatty ordered on approval Coptic materials offered by Tano for £ 235. That year Tano wrote Beatty from New York not to involve his American-based nephew Frank J. Tano in any transaction, but to remit directly to the Ottoman or Barckley Banks of Famagusta, Cyprus. On 12 September 1951 Tano wrote Beatty's secretary John Marsh in London:

> I asked to [*sic*] a friend in Paris to forward threw [*sic*] you for Mr. Chester Beatty a collection of Coptic parchemains [*sic*]. Please wen [*sic*] you receive them, kindly forward the parcel to Mr. Chester Beatty's address.

On 25 iii 54 Beatty's secretary John Wooderson recorded in a memorandum:

> Mr. A. Chester Beatty asked John Wooderson to see Mr. Tano and find out if he had any Coptic writing or vellum or pages of papyri in Greek; and if so, what they would cost, and if they could be examined in London.... Mr. Tano said he had no stock in Cairo or Cyprus at present but that he would write later if he found anything interesting.

Thus business relationships were well established long before the mid-'50s, when Tano did find much that was very interesting, which he proceeded to sell partly to Martin Bodmer and partly to Sir Chester Beatty.

Tano kept his collection in part in Cyprus, as he was able to get it out of Egypt. He could correspond with Beatty freely from Cyprus while there for the summers, and could send the material and receive payment without difficulty. Tano was from time to time on the continent, and Beatty seems to have had direct contact with Tano in Nice. This relationship outside of Egypt was not only convenient from the point of view of customs and payments, but was also diplomatically advantageous, as is reflected in a comment of Beatty to Wilkinson's successor as Librarian, Dr. R.J. Hayes (Director of the National Library of Ireland from 1940 to 1967), in a letter of 21 March 1958 concerning ac. 2554: "We can honestly say it was bought in Europe; we need not say where or when".

On 2 April 1956 Beatty wrote Wilkinson from Nice that he had seen Tano and "got some very interesting things from him". In a letter of 5

April 1956 to Wilfred Merton, his papyrological consultant in Dublin, Beatty was more specific, distinguishing the following items clearly enough for us to identify them, in the light of later information:

> The two books with the original bindings are very interesting. One seems to be complete and the other was never finished. About half of the papyrus pages are blank. [Ac. 1389 and ac. 1499].
>
> It was evidently a scroll which was cut in pieces to make it appear like a book. It must have been pretty long, because it is quite thick — it must be 2" at least — and the page is just the size of the section of a scroll. They just bend over, and I looked at a good many of the pages and they separate naturally, so I do not anticipate much trouble in having the proper experts separate them. [Ac. 2554].

Beatty added: 'I will, of course, deliver them at once to the British Museum when I arrive".

On 15 April 1956 Wilkinson replied urging Beatty to invite the leading authority on book bindings, Berthe van Regemorter, to come from Belgium to examine the bindings before the books were disassembled and the leaves glassed, a proposal with which Beatty readily agreed.

Mlle Van Regemorter had recently been at the Bibliothèque Bodmer to examine the book bindings there, and had sent Beatty a report concerning her findings. The ensuing discussion illustrates the way in which Bodmer and Beatty became involved in friendly competition for Tano's wares. For in a letter to Merton of 21 May 1956 Beatty commented:

> You have seen the memorandum that Miss Van Regemorter did on Bodmer's library. Apparently he got some good things from Tano. It was quite an important purchase, and I imagine it was the Gospel of St. John that he bought. I do not think he is making a general collection of papyri, but I think he bought a few very important things from Tano.

In a letter the same day to Wilkinson Beatty reported that Tano had confirmed the sale to Bodmer:

> He indicated to me that he had an important deal on with Bodmer. I imagine it is in connection with that Gospels. Anyhow, I hope we will get some other things, and I wrote to him about early wooden bindings. I imagine Bodmer is not going in for those, and he [Tano] may be able to clean up the market and get something fine there.

Actually, Bodmer had been in Cairo late January to early February 1956, after having alerted Father L. Doutreleau, S.J., a patristic scholar of the French series Sources Chrétiennes, at the time posted in Cairo, to assess what Tano's "Bey of Papyrus" had available. P. Bodmer XIV-XV (P 75 = the Gospels of Luke and John) reached Geneva in February 1956. Indeed a number of important Dishna Papers reached Geneva in 1956, such as three lost plays of Menander (P. Bodmer XXV, IV, XXVI). This tends to suggest that what was offered to Beatty in the Spring of 1956 may have been things that Doutreleau and/or Bodmer considered not to be "world literature", as Bodmer described his scope[27], but rather the kind of artifacts, such as book bindings, that interested Beatty.

In a letter of 24 April 1956 to Merton, Beatty described quite candidly his business procedure with Tano:

> You see, with the deal I had with "X" [Tano], I pay so much for the whole lot, and if I do not want to buy the whole lot I pay another sum. I pay £ 800 if I take them all, but if I do not take the whole lot I pay £ 200, but I can pay this in sterling. In other words, the price was 10,000 Swiss francs, which is a little over £ 800, and it is done on the normal exchange basis. Then I have the right to pay in sterling. Of course, it is a good deal like buying a pig in a poke, because he does not know too much about them and I know nothing. They look old and they smell old, and I imagine they are old. That is the opinion of a real expert.

A report about Tano in September 1956 on file in the Cologne archives gives insight into the transactions from Tano's point of view:

> He is 60 years old, one of the most serious antiquities dealers in Egypt. Yet his business really blossomed only after the death of Nuʿmān [Beatty's previous agent in Egypt, through whom e.g. the Manichaean codices of Medinet Madi were acquired]. Tano had been invited by Bodmer and Chester Beatty to Europe (Geneva and Nice), and had flown for the first time in his life. Having just arrived back, he was still so overwhelmed by the impression that he narrated the whole course of his trip to me — which according to the orient's rule of secrecy he would no longer have done a week later. First in this moment did I realize what path the great papyrus discoveries take today: Tano is the permanent agent of Bodmer and Chester Beatty. He is advised by ... Dr. Wild in the Institut Français. Everyone knows that he is constantly buying and sending to Europe interesting papyri for these two clients. Also Nisim Cohen [a rug dealer who had also sold manuscripts to Bodmer] has already made deliveries to Tano. Tano sends the texts to his clients even on approval to Europe.

Beatty lacked the expertise provided to Bodmer in Cairo by Father Doutreleau, and was dependent on expertise he received once he had

[27] Martin Bodmer, *Eine Bibliothek der Weltliteratur* (Zürich: Artemis, 1947).

taken an option to buy and had directed the material to the British Museum.

Beatty was also at a disadvantage in terms of convertible funds. In a letter of 26 April 1957 to Mlle Van Regemorter he explained:

> You see, he [Bodmer] buys from the same man who offered me the things. He is a dealer in Cairo and a very old friend of mine. I am rather handicapped, as Bodmer can pay in Swiss francs for everything he buys, and I do not like to ask the Bank of Ireland for foreign currency, except on very rare occasions.

A chance late in 1956 offered by Mlle Odile Bongard, Bodmer's secretary, was such a rare occasion, as Beatty wrote to Wilkinson on 8 January 1957:

> Confidentially I have some very good news. We have got some more things from Tano, and it looks as if they are very good. I am just arranging to buy them and have asked Miss O'Neill [Business Secretary in Dublin, 1950-68] to take it up with the Bank of Ireland and get the money. I have to pay for them in Swiss francs, and they came to 4,000 Swiss francs, which is about £ 360 to £ 375.

In his letter of 21 May 1956 to Merton concerning the first acquisition, Beatty wrote how he planned to reach a decision as to whether to exercise his option:

> My idea is, soon after I arrive, to take the big papyrus which is cut apart [ac. 2554] and in that parcel there are two lots of loose leaves — one is supposed to be agnostic [?] — and have them identified at the British Museum. I will not do anything beyond identification, because I do not want to be forced to take the lot, in case the other two are of no value.
>
> ...If she [Mlle Van Regemorter] can come over we will take the other two books [ac. 1389 and ac. 1499] to Edwards and Skeat at the British Museum [so] she can study the bindings. In the meantime, we will have the option [for: opinion?] about the first lot.

Thus in addition to ac. 1389, ac. 1499 and ac. 2554, "there are two lots of loose leaves", or, as Beatty wrote Wilkinson on the same date, "two other odd lots of text". This may be correlated with the comment in the Registry of Accessions at 1389 that it was acquired "summer 1956 with ac. 1390 and 2 boxes of loose leaves". One may infer that ac. 1390 was originally included with the "two lots", but on its identification was distinguished from the rest of the contents of the two boxes.

Mlle Van Regemorter's brief report on ac. 1389, ac. 1499 and ac. 2554 is on file at the Chester Beatty Library, as is that of T.C. Skeat, a classical scholar in the Department of Egyptian Antiquities of the British Museum, dated 4 June 1956. It is obvious that the decision was made to acquire the material Tano offered in the Spring of 1956.

In 5 July 1956 I.E.S. Edwards, an administrator in the Department of Egyptian Antiquities, wrote to Beatty in Dublin:

> I am sending you some photographs of the two books which we made for record purposes. [Stanley] Baker [the conservator of papyrus] has already placed all the loose leaves of the Coptic manuscript [ac. 1389] between glass, but many fragments have yet to be placed. As one would expect they are in some disorder, but it will not be difficult to remedy that defect. Ultimately, I hope to be able to send you a description noting any passages which may be missing.
>
> ...Work on the Greek manuscript [ac. 2554] has hitherto been limited to the removal of the outermost loose leaves, which we shall ask Skeat to study in due course.

The accompanying photographs are clearly identifiable as ac. 1389 and ac. 2554, which confirms the identifications made tentatively on the basis of the descriptive phrases in the letters.

On 10 July 1956 Edwards wrote Beatty details about the contents of ac. 1389, "the Coptic manuscript":

> As far as my examination has gone (to p. 58), it is all Joshua; probably Judith follows, as the order on the title page suggests.

Thus ac. 1389 is identified as the cover and over half the leaves of a codex of Joshua — in spite of the reference to Judith on the title page, the text of Judith is not present. The rest of the codex was acquired by Bodmer and published together with ac. 1389 as P. Bodmer XXI.

Skeat's report, after listing ac. 1499 and ac. 2554, commented:

> In addition to the above are two small folders of papyrus fragments. One of these, marked X, contains a small fragment of a leaf of a papyrus codex of the 3rd cent. A.D., containing parts of John xix. 26-27, 31-32; this may be from the John codex recently acquired by M. Bodmer [P. Bodmer II], and I should be grateful for permission to inform M. Bodmer of its whereabouts. Some of the other fragments come from Codex B above [ac. 2554]. There are also 3 or 4 small fragments of what appears to be a Gospel narrative (late 2nd cent. A.D.), and some Coptic pieces [presumably including ac. 1389 and ac. 1390].

The presence of the fragment of P. Bodmer II makes it clear that the materials acquired from Tano came at least in part from the same discovery as did the Bodmer Papyri. To be sure, the fragments of a Gospel narrative and the Coptic pieces are too imprecisely described to be readily identifiable. The former were no doubt in Greek, since they might otherwise have been subsumed under the Coptic pieces, but have apparently been subsequently lost from sight. In fact Skeat's identifications seem to have been limited to Greek material. The Coptologist

A.F. Shore was soon added to the staff of the Department of Egyptian Antiquities.

On 23 xii 58 Skeat wrote Beatty:

> You will recall that, with your permission, I sent Prof. Martin of Geneva, who is editor of the Bodmer papyri, a photograph of the small fragment of St. John's Gospel which I noticed in your last purchase. He has just written to say that it is undoubtedly part of the codex bought by Bodmer, and in fact bridges the gap between two detached fragments in the Bodmer collection, making it possible to reconstruct the entire page. At the same time, further discoveries have been made in the placing of small fragments of the codex in the Bodmer collection, and Prof. Martin is considering a further publication which would bring these to the notice of scholars. He has asked me to ascertain whether you would be able to agree to the text of your fragment being published together with that of the Bodmer fragments to which it joins; and in passing this request to you, I would venture to say that I hope you will agree.

On 29 xii 58 Beatty wrote Skeat:

> As regards the fragment for Dr. Bodmer, I should be very glad for him to reproduce any of the fragments that I have in any way he sees fit. I am glad that I have something which joins up some of his pieces. I do not mind about the fact that it may be used by us in some of our publications.

Neither the *Supplement* of 1958 nor that of 1962 to *Papyrus Bodmer II* (1956) included the fragment at the Chester Beatty Library in the transcription, but photographs are included on plates 135/136 of the 1962 *Supplement*[28].

The fragement of P. Bodmer XX glassed with that of P. Bodmer II

[28] *Papyrus Bodmer II. Évangile de Jean chap. 1-14*, ed. Victor MARTIN, Bibliotheca Bodmeriana 5 (Cologny-Geneva: Bibliothèque Bodmer, 1956). *Papyrus Bodmer II. Supplément. Évangile de Jean chap. 14-21*, ed. Victor MARTIN (Cologny-Geneva: Bibliothèque Bodmer, 1958). *Papyrus Bodmer II. Supplément. Évangile de Jean chap. 14-21. Nouvelle édition augmentée et corrigée avec reproduction photographique complète du manuscrit (chap. 1-21)*, ed. Victor MARTIN and J.W.B. BARNS (Cologny-Geneva: Bibliothèque Bodmer, 1962). Kurt TREU, "Christliche Papyri 1940-1967", *Archiv für Papyrusforschung* 20 (1970) 182, correctly lists the passage as John 19:25-28/31f. Kurt ALAND, ed., *Repertorium der griechischen christlichen Papyri 1: Biblische Papyri: Altes Testament, Neues Testament, Varia, Apokryphen* (Patristische Texte und Studien 18; Berlin and New York: Walter de Gruyter, 1976) 296, lists the fragment under folio 135/136, which is the numeration of the plates but not that of the (restored) pagination 139/140 in Papyrus Bodmer II, *Supplément, Nouvelle édition*. Similarly Joseph van HAELST, *Catalogue des papyrus littéraires juifs et chrétiens* (Université de Paris IV Paris-Sorbonne; Série "Papyrologie" 1; Paris: La Sorbonne, 1976) 156, entry 426, lists it as "fragment of a leaf paginated [*sic*!] 135-136". The divergence between the plate numbers and the postulated pagination is due to the fact that pp. 35-38 are missing, with the result that the numeration of the plates became four digits lower than the missing pagination.

(and hence sharing an accession number 2555) seems to be another of Skeat's identifications, though not mentioned in his report. For according to the *editio princeps* it "has been discovered by Mr. T.C. Skeat among the papyri of the Chester Beatty collection. We are indebted to his courtesy for the communication of the text of this fragment that happily completes this page"[29]. At the Bibliothèque Bodmer photographs of the Chester Beatty fragment are glassed with this Bodmer leaf in the lacuna. Thus, Skeat presumably communicated to Bodmer the existence of this fragment as well as that of P. Bodmer II, perhaps in that the photographs he sent included the three fragments that were glassed together. The small fragment of P. Bodmer XX may not have been initially identifiable by Skeat in 1956, though the identification did take place before P. Bodmer XX was published (1964).

On 7 i 57 Beatty wrote from Nice to Merton of a second potential acquisition:

> I received a letter from Bodmer's secretary [Mlle Bongard] who had just come from Cairo, as he had sent her to go through all the fragments that Tano had in the hope of finding a few little fragments which had been overlooked of the St. John's Gospel, and she managed to find a few fragments. She told me that she had certain things which Tano wanted me to have, and she told me the price was 4,000 Swiss francs, and I asked her if she would leave them with me, as I wanted to get a little information on them, and I would probably take them. There are 8 items, of which 6 are papyrus, and one, curiously enough, a perfect mass of small fragments. In fact, they fill a small plastic box of about 4" long by 3 ½" wide by 2" deep. Then there is a roll on vellum of some sermon [ac. 1486], which is quite early. Some things are supposed to be of the 5th century and others of the 9th century .... So when Lady Powerscourt went back, I sent samples of the find, with the exception of one item, to Edwards.

[29] *Papyrus Bodmer XX. Apologie de Phileas évêque de Thmouis*, ed. Victor Martin, with an *Essai de reconstruction du texte original grec* bound separately (Cologny-Geneva; Bibliothèque Bodmer, 1964), p. 34. Eric G. TURNER, *Greek Papyri: An Introduction* (Oxford: Clarendon, 1968, supplemented paperback edition 1980) 52, states that "a few scraps were found among Sir Chester Beatty's texts", and van HAELST, *Catalogue des papyrus littéraires juifs ert chrétiens* 258, item 710, reports "tiny fragments at Dublin". *Papyrus Bodmer XX* lists only one such fragment with acknowledgement to T.C. Skeat for identifying it, pp. 34, 36. This fragment at the Chester Beatty Library can be distinguished in the *editio princeps* by the darker coloration of its photographs on plates 7 (135, 13-16) and 8 (136, 14-17). A cut-out of the fragment from a photograph is glassed with the rest of the leaf at the Bibliothèque Bodmer. It has been republished by A. PIETERSMA, "New Greek Fragments of Biblical Manuscripts in the Chester Beatty Library", p. 47 and Plates 9-10. The fact that the fragment itself is glassed in Dublin with two others, one from P. Bodmer II, the other unidentified (but not from P. Bodmer XX), may have led to the plural. PIETERSMA, p. 48 and Plates 9-10, identifies the hand of the unidentified fragment with that of Chester Beatty Biblical Papyrus VI.

On 16 xii 56 Beatty had written to Edwards a letter following up his shipment of samples:

> I am sending you a package containing some papyri, which I have labelled A, B, C, D, E, F, G and H, i.e. eight packages. Actually, there are six papyri, some of which are very fragmentary, and 'E', which is in a little tin box, is on vellum and apparently a Coptic sermon [ac. 1486]. 'H', which might be the most interesting, is a Syriac MS. It has no miniatures, and I do not know whether it is complete, but it has 75 pages and is in good condition.
>
> I should be very pleased if you would get the proper advice and find out if they are of any value. I do not know what to make of these fragments. One lot they say is from the same roll as the Greek papyrus we have of the time of Diocletian [ac. 2554], and there are two big lots of fragments which are still here and I will get to you later.

On 21 i 57 Edwards wrote Wilkinson that "the latest Coptic documents ... seem to me to be too fragmentary to be very promising". On 13 ii 57 he wrote Beatty in more detail:

> Perhaps I can explain the present position most clearly if I say that we appear to be dealing with nine different documents, all fragmentary. The plastic box, which we had opened before I received your letter, proved to be a kind of pool of fragments and we have already drawn on its contents for building up parts of nearly every one of the nine documents. You would, I think, have been surprised to see what a state of confusion these fragments were in: there was not one packet which did not contain elements of more than one document.

The same letter included a comment about ac. 1390 from the first acquisition, although it is not clear if fragments from the more recent acquisition are involved:

> In the meantime I can only report that more of St. John has been joined up; eventually I think it may prove to be quite a useful find.

The deaths of Wilkinson (28 i 57) and Merton (2 xi 57) must have disrupted the oral tradition at the Chester Beatty Library. Miss Evanna McGilligan, apparently relatively new on the Library staff, wrote Edwards on 5 ii 58:

> As I have no report here of the papyrus books of John [ac. 1390], Joshua and Judith [ac. 1389] and the Greek texts of a school exercise [also ac. 1390 — the Registry of Accessions at 1390 distinguishes "one separate glass of Greek text"], I wonder if you could be so kind as to give me a brief note on when these things were acquired and where, if this information is available. Some description of these things, however short, would be of great help to me.

On 6 ii 58 Edwards replied:

> They were bought by Sir Chester from a dealer in Cairo [appended

> handwritten note: I know privately that he was Tano...] in 1956. We know nothing about their provenance and there is nothing in their contents to suggest its location. They all date from about the IVth century A.D.
>
> You may have heard that Sir Chester has recently sent us some more fragments of the texts of St. John [ac. 1390]. If sometime the sheets in Dublin could be brought over to this Museum the new fragments could be joined to them.

From a letter of Miss McGilligan to Beatty of 11 ii 58 it is clear that ac. 1390 had been "brought to Dublin by him [Edwards] last year". This would indicate that ac. 1390 was transferred to Dublin after the placements mentioned on 13 ii 57 had been compled, no doubt on the assumption that all the fragments that were extant had been placed. On 21 ii 58 Miss McGillighan wrote Beatty again:

> Miss Callaghan left Dublin this morning and brought with her the papyri of the Coptic St. John which Mr. Edwards is going to work on at the Museum. He is going to join the recently acquired fragments to the leaves already mounted last summer.

Since there are no indications that a further purchase had been made later in 1957, one may conjecture that the reference to "the recently acquired fragments" means that they were only recently received at the British Museum. For in his letter of 7 i 57 Beatty mentioned to Merton that he had only "sent samples of the find" to Edwards, indeed that "there are two big lots of fragments which are still here and I will get to you later". Perhaps the remainder was only belatedly sent, after the conserved leaves had been transferred from London to Dublin.

On 15 iv 58 Beatty wrote to Hayes:

> I have just to-day received a most interesting letter from Mr Edwards. He says: "The fragments received a few months ago have practically all been placed. In interest and quality they are certainly a truly remarkable little collection. Together with those which were previously in your possession they must rank as one of the three or four best collections in the world to-day". This remark applies to our Coptic papyri, so that gives the library quite a boost, I think. I was delighted to get his opinion — he is a most conservative man — and I knew you would be too.

There was a still further acquisition made at this time in 1958, again mediated through Mlle Bongard of the Bibliothèque Bodmer. On 18 xii 57 Tano wrote to Beatty in Nice:

> I wrote to Miss Odile Bongard to forward you some papyrus which completes some you bought before. Also if she received a lot of parchements in Coptic. In case she did, please their price send it if possible in Cyprus pounds.

Miss McGillighan wrote 10 iv 58:

> I will be very pleased to go to Geneva and collect the papyrus from Mademoiselle Bongard, as you suggest. I had planned to leave Paris for Dublin on May the 18th and so it would be on May the 19th that I would go to the Bodmer Library and collect the papyrus.

On 19 iv 58 Bodmer wrote Beatty: "The package from Tano is also ready to be delivered to you!" The package seems to be an item distinct from the papyrus fragments, and presumably contained the "lot of parchments in Coptic".

On 23 v 58 Miss McGillighan wrote Beatty:

> I collected the package which contains some fragmentary leather bindings and 17 vellum folios with some fragments, one with a miniature, several with spiral ornamentation and several with coloured initials. They are in fairly good condition and Dr. Hayes thinks that the writing may be Greek, but I would opt for Coptic. Dr. Hayes believes that it may be a find comparable to the Syriac MS.

The papyrus fragments in this acquisition may well be what "completes" ac. 1390, even though in this report limited to the package there is no explicit reference to the papyrus.

From the photographs of ac. 1390 made at the British Museum, one can clearly distinguish two stages in conservation, presumably one in 1956-57 and one early in 1958. A.F. Shore's report of vii 58 must, to judge by its date, presuppose both stages. But it lists the text of ac. 1390 as going only as far as John 12:14, where the tenth page (the seventh from the Gospel of John) ended. The next leaf extant today (John 12:14-27/27-39) underwent three stages of photography, and only at the last stage is the largest fragment, representing most of the leaf, present. A red book of photographs at the British Museum reported that the third series of this leaf was made "after repair July 1962". A further relatively complete leaf and two final leaves consisting only of three fragments each concluding at John 13:38 are extant today. Thus between vii 58 and vii 62 there seems to have been a still further conservation, perhaps involving material picked up in Geneva 19 v 58.

Tano's business dealings with Beatty continued, for even as late as 11 v 59 Beatty wrote to Hayes:

> I have some very confidential and interesting news for you. I finally persuaded a friend of ours in Egypt to visit me, and I am now on the track of getting two more interesting things. One is a 4th century Coptic manuscript on papyrus — something biblical. It is of course very early, and I do not know of any of these Coptic books on papyrus, only the scrolls. The other thing is a long Book of the Dead, about 16 ft. long. I do not know whether it is in colour. If it is in

> colour it will be particularly interesting, and I do hope it is. I am hoping they can be bought at a reasonable price.

Although the Book of the Dead is presumably not from the same discovery, the Coptic papyrus biblical codex of the Fourth Century might well be. It is not known whether this acquisition was actually made; it has not been identified as such at the Chester Beatty Library.

Sir Chester Beatty thus seems to have made a series of acquisitions from Tano during 1956 and the time immediately following. At crucial junctions Bodmer seems to have had first pick, for one wonders if shipments made from Tano to Beatty via Bodmer may not to some extent have been items that Bodmer had declined to buy. In the case of the second shipment, consisting mostly of fragments, it would seem clear that Mlle Bongard had sorted through the lots and removed any she identified as belonging to P. Bodmer II before passing on the residue to Beatty. But even the first acquisition in iv 1956 may have involved items not recommended by Doutreleau and bypassed by Bodmer when he visited Tano in Cairo just prior to his meeting with Beatty in Nice. The largely uninscribed codex (ac. 1499) and the unfinished one (ac. 2554), also largely uninscribed, would both be the kind of items that Bodmer would not acquire. Only in the case of ac. 1389, half of P. Bodmer XXI = the Coptic Joshua, can one assume Bodmer would have been interested in it, unless indeed the leaves he did acquire from that codex were merely part of a lot that he had acquired for other reasons.

Much of the present task has been to identify the various vague references from the letters in terms of identifications that could only be made once the material was at the British Museum to be glassed, identifications then passed on in typescript to the Chester Beaty Library and preserved there, in archival documents and in the Registry of Accessions. The only item with even a partial identification before being sent to the British Museum was "apparently a Coptic sermon" "on vellum". This must be ac. 1486, the Pachomian Abbot Theodore's Letter 2, item V in Shore's report of vii 58 (also quoted in the Registry of Accessions), where it is referred to as "apparently a letter or sermon of reproof".

Three of the items here under discussion have been identified without further ado as stemming from the same discovery as do the Bodmer Papyri, simply because they are pieces of the same documents, P. Bodmer II (the Gospel of John, ac. 2555), P. Bodmer XX (the *Apology of Phileas*, Bishop of Thmouis, ac. 2555) and P. Bodmer XXI (Joshua, ac. 1389).

In addition to these, ac. 1390 published here has been clearly identified with the same discovery by the slip of paper stapled in the Registry of Accessions and by the references to this papyrus in the correspondence reported here.

A good number of other documents from the same acquisitions and hence presumably at least in part from the same discovery are know from the correspondence to have been acquired by the Chester Beatty Library. In some instances the texts can be identified by procedures similar to that employed in the case of ac. 1390, though with diminishing degrees of certainty. Those listed in the inventory are those with a reasonable probability of coming from the same discovery, though one cannot reach final certainty in some instances; and others not included here, for lack of adequate information, may nonetheless be from among the Dishna Papers.

3. A Codicological Analysis of Accession 1390

According to the Registry of Accessions, ac. 1390 is conserved in "five glasses". There is reference to "also one separate glass of Greek text of a school exercise and a leaf cojoined to the text of St. John".

The total of six glasses in the Chester Beatty Library contains eight leaves, since in two cases conjugate leaves are in the same glass (extant leaves 6 and 7 and 5 and 8 [original Coptic pagination has survived nowhere]). The eight leaves, with the identification on the labels of each glass pane in quotation marks, are as follows:

> Leaf 1 = extant pp. 1/2: ↑ (vertical fibres) "Greek text (School exercise)" / → (horizontal fibres) "Greek text (School exercise)". This leaf, from the first half of the quire, is the third leaf from the center of the quire, and is conjugate with leaf 6. It is 25.4 cm high and 14.0 cm broad.
>
> Leaf 2 = extant pp. 3/4: ↑ "Greek text (School exercise)" / → [the conclusion of the Greek text and:] "Continuation of Gospel of John. Subachmimic. 1. John x 8 [7]-18". This leaf, from the first half of the quire, is the second leaf from the center of the quire, and is conjugate with leaf 5. It is 25.5 cm high and 13.9 cm broad. There is a kollesis visible on the side of leaf 2 with horizontal fibres, about a third of the distance in from the left edge of the leaf, with the kollema on the right overlapping that on the left.
>
> Leaf 3 = extant pp. 5/6: ↑ "2. John x 18-36" / → "3. John x 36-XI 12". This leaf, from the first half of the quire, is at the center of the quire, and is conjugate with leaf 4. It is 25.2 cm high and 13.9 cm

broad. In the bottom margin of p. 5 the Coptic numeral 177 occurs, without its significance being clear. It could hardly be the line count of five pages with approximately 35 lines each, the number of lines on this page, since in the present case the text of John had only begun on p. 4 where 22 lines were copied; the preceding mathematical exercises average much less than 35 lines per page.

Leaf 4 = extant pp. 7/8: → "4. John xi 12-38" / ↑ "5. John xi 28-43". This leaf, from the second half of the quire, is at the center of the quire, and is conjugate with leaf 3, with which it is conserved. It is 25.5 cm high and 13.9 cm broad, producing with leaf 3 a breadth of 27.8 cm for the whole sheet. It is identifiable as the center of the quire by the fact that the text moves from the horizontal side of leaf 3 to the horizontal side of leaf 4 without a break. Since the sides with horizontal fibres face each other when the quire is closed, the quire is of the usual kind, where horizontal fibres face inward and vertical fibres face outwards in the closed quire. There is a kollesis visible on the side of leaf 4 with horizontal fibres, at the inner margin of extant p. 7, with the kollema on the right overlapping that on the left.

Leaf 5 = extant pp. 9/10: → "6. John xi 43-55" / ↑ "7. John xi 55 [56]-XII 14". This leaf, from the second half of the quire, the second leaf from the center of the quire, is conjugate with leaf 2, with which it is conserved. It is 25.3 cm high and 14.2 cm broad, producing with leaf 2 a breadth of 28.1 for the whole sheet.

Leaf 6 = extant pp. 11/12: → "8. John xii 14-27" / ↑ '9. John xii 27-39". This leaf, from the second half of the quire, the third leaf from the center of the quire, is conjugate with leaf 1. Yet they are not conserved together, perhaps because leaf 1 contains only the Greek school exercise and leaf 6 only material from the Gospel; they may not have been originally identified as part of the same quire. Leaf 6 was also reconserved late in the procedure. It is 25.2 cm high and 14.4 cm broad, producing with leaf 1 a breadth between 28.4 cm, the sum of the two breadths, and 24.0 cm, the breadth when the two leaves are touched near the bottom of the sheet. There is a kollesis visible on the side of leaf 6 with horizontal fibres, at the inner margin of extant p. 11, with the kollema on the right overlapping that on the left.

Leaf 7 = extant pp. 13/14: → "10. John xii 40 [39]-XIII 2" / ↑ "11. John xiii 2-12". This leaf, from the second half of the quire, is the fourth from the center of the quire; its conjugate leaf is not extant. It is 23.4 cm high and 13.8 cm broad. There is a kollesis visible on the side of leaf 7 with horizontal fibres, at the inner margin of extant p. 13, with the kollema on the right overlapping that on the left.

Leaf 8 = extant pp. 15/16: → "12. John xiii [13]-14...[17]-18...23 [22]-6" / ↑ "13. John xiii 28 [27]...31-[33]...35-8". This leaf (actually three

> fragments of a leaf, one above the other), from the second half of the quire, is the fifth from the center of the quire; its conjugate leaf is not extant. As conserved, the extant parts of the leaf are positioned to produce a total height of 19.1 cm; the third, broadest fragment, is 5.5 cm broad.
>
> One may conclude that the leaves when relatively intact were 25.2 to 25.5 cm high and 13.8 to 14.2 cm broad, with a breath for a sheet near the center of the quire of 27.8 to 28.1 cm broad. There is continuity of horizontal fibres from the left edge of sheet 1+6 to the right edge of sheet 2+5 and from the left edge of sheet 2+5 to the right edge of sheet 3+4. This would indicate that the roll was cut from right to left and the sheets stacked successively one on the other to produce the quire. Since there is a kollema on each of the three completely extant sheets, the breadth of a kollema must have been not much different from that of a sheet. Since the leaf conjugate to leaf 7 is missing, whose left edge should have joined the right edge of sheet 1+6, it is difficult to determine whether leaf 7 belonged to the same roll, although continuity of horizontal fibres seems possible (but not certain) from the left edge of leaf 7 to the right edge of sheet 1+6; a kollesis on the missing leaf would have produced a different fibre to align with sheet 1+6. Nothing can be said in terms of continuity of horizontal fibres from the left edge of leaf 8 to the right edge of leaf 7, where again a kollesis on the missing leaf would have produced a different fibre pattern to align with leaf 7. Thus it is not certain whether the 5 sheets represented in the extant material derive from the same roll, representing a total breadth of some 140 cm, or only the top or central 3 sheets, representing a total breadth of some 84 cm. There is no positive evidence of there having been more than one roll or more than one quire, but neither possibility can be excluded.

Shore's typed report of vii 58 on file at the Chester Beatty Library states regarding this codex that there are "12 leaves or fragments of leaves (6 folios and 3 bifolios)", whereas, as the inventory above makes clear, in fact there are only 2 single folios (leaves) and 3 bifolios (sheets), a total of 8 leaves. The 3 bifolios are the sheet at the center of the quire (leaves 3+4 = extant pp. 5/6+7/8); the second sheet from the center of the quire (leaves 2+5 = extant pp. 3/4+9/10); and the third sheet from the center of the quire (leaves 1+6 = extant pp. 1/2+11/12). Only 2 folios remain, both from the second half of the quire: the fourth leaf from the center of the quire, leaf 7 (extant pp. 13/14); and the fifth leaf from the center of the quire, leaf 8 (extant pp. 15/16). The other 4 leaves mentioned by Shore (of which old photographs are not available) must be in fact "fragments of leaves". For they are no longer conserved as separate leaves, but presumably are conserved as placed fragments on the 8 leaves that are extant. Thus Shore's report of vii 58 does not reflect the present state of conservation, but a preliminary stage later rectified and supplemented[30].

Shore's report remarks regarding the two leaves with the Greek text (leaves 1-2 = extant pp. 1/2-3/4): "It is uncertain whether these leaves are continuous". Indeed one mathematical exercise concludes at the end of the one leaf and another begins at the beginning of the other. But once one has confirmed the continuity of the text of the Gospel of John from the leaves conjugate with leaves 1 and 2, namely from leaves 5 to 6, then the continuity of text from one school exercise to the next can be assumed, since no leaves would intervene between leaves 1 and 2. This assumption that no intervening leaf is missing is confirmed codicologically by the continuity of horizontal fibres from one sheet to the next: from the right edge of the sheet consisting of leaves 2+5 (extant pp. 4+9, the horizontal-fibred side of the sheet next to that which is at the center of the quire), on to the left edge of sheet 1+6 (extant pp. 2+11, the horizontal-fibred side of the next sheet, one sheet further from the center of the quire).

[30] Tito ORLANDI, "Les Manuscrits coptes de Dublin, du British Museum et de Vienne", p. 326, speaks of "12 leaves", and adds "on the *recto*, mathematical exercises in Greek". This could not be an inventory of the material as actually glassed and available for examination at the Chester Beatty Library. It is presumably an abbreviation of Shore's report (though by omitting "or fragments of leaves" it falsifies it). The abbreviated use of Shore's report is also suggested by the ambiguous use of Shore's term *recto*. In Shore's report *recto* means the first side inscribed, not the side with horizontal fibers, as used to be customary, for in the present case the side he called *recto* actually has vertical fibers. Shore's terminology is made clearer by his addition, omitted by Orlandi, "and half the *verso*", where the mathematical exercises end. Thus it is clear that the side with vertical fibres was inscribed in Greek first, then this Greek text completed at the top of the side with horizontal fibres, then the Coptic text begun on the rest of the side with horizontal fibres.

MATHEMATICAL SCHOOL EXERCISES IN GREEK

WILLIAM BRASHEAR

Introduction

Leaves 1 and 2 (= extant pp. 1/2, 3/4) of the Chester Beatty codex Ac. 1390 contain a series of exercises in arithmetic and plane and solid geometry written in Greek. The fluid, cursive script is typical for documents of the mid-third to mid-fourth centuries A.D. and can be compared with other examples from this time-span, e.g. M. Manfredi Trenta Testi Greci da Papiri Letterari e documentari editi in occasione del XVII Congresso Internazionale di Papirologia, Florence 1983, no. 28; *Greek Papyri in the British Museum. Facsimiles*, vol. III, London 1907, no. 748; R. Pintaudi, *Dai Papiri della Biblioteca Medicea Laurenziana II*, Florence 1977, no. 24; W. Schubart, *Papyri Graecae Berolinenses*, Bonn 1911, pl. 34b; R. Pintaudi, ed., *Papyrologica Florentina* XII: *Papiri greci e latini a Firenze*, Florence 1983, pl. XXXVII.

In addition to the paleography other criteria for dating these texts are: 1) the presence of calculations involving myriads of denarii. Such inflationary amounts are common in papyrus documents from the fourth century onwards. However, their appearance here may be merely haphazard and have nothing to do with contemporary economics - cf. the multiplication exercises in *ZPE* 14 (1974) 61-66. 2) The letters topped by a curl as thousand-indicator. In the mid-fourth century this usage was abandoned in favor of a diagonal stroke below and to the right, the left or the center of the cipher (Brashear, *ZPE* 60 [1985] 239f.), thus marking the water-shed in dating documents with thousands-numbers in them. Since the change-over to ͵A, ͵B and ͵ϛ has evidently not yet taken place in these texts, it is tempting to date them before the mid-fourth century A.D.

Individual letters vary considerably in size and cursiveness — compare and contrast p. 2 lines 12 and 13 with line 16, for example; or the different shapes of the etas on p. 2 line 24 and p. 3 lines 24, 25, 27. Large, quasi-ornamental letters sometimes adorn the beginnings or ends of lines, e.g. p. 2 line 15: πόσων, l. 20: ταῦτα, p. 3 l. 25: κράτησον. However, they mark the beginnings neither of new problems nor even of new sentences and seem to be placed entirely at the whim of the writer. Likewise, the large letters at line-ends could conceivably be regarded as space-fillers. However, numerous shorter lines have quite normal-sized concluding letters, leaving large spaces gaping between the last word and the edge of the papyrus.

Problems are separated by a variety of devices: by indenting the concluding line of a problem (p. 2 l.6), by a horizontal line (p. 3), or by an ornamental mark (p. 3 l.24). Sometimes there is no division whatsoever (p. 3 l.21), and one problem follows directly upon another.

The only diacritical marks are an apostrophe in the phrase τουτ' ἔστιν (*passim*) and a trema over the iota in ἴσος (p. 3 line 7).

For the most part the orthography and grammar are correct. The following types of errors may be noted[1]:

haplography of a letter (p. 3 l.3); of a whole line (p. 2 l.17-18).
dittography: πολυπλασίασον (p. 1 l.7-8).
incorrect case: κάλαμοι (p. 3 l. 3); κάλαμοι (p. 3 ll.13,14).
incorrect gender: ποίῳ ψήφῳ (p. 3 l.20).
incorrect mood: τρίπλωσον (p. 3 last line).
incorrect vowel length: τον (p. 1 l.10); τουτον (p. 4 l. 4); πόσον ναύβιον (p. 1 ll.3-4).
misspellings: κατησον (p. 3 l.21); στρογυλουν (p. 1 l.15); εφροχου (p. 2 l. 7).

Whole numbers are either feminine or neuter in gender (cf. *MPER* n.S. 1, p. 48); fractions neuter - both normal usages in mathematical texts.

Whole numbers and fractions alike are usually capped by a horizontal stroke. Only rarely is the stroke missing. Occasionally the fraction is indicated by two diagonal dashes to the right of the cipher (p. 3 lines 19ff.: γ''; p. 1, last line: ι''). One-half is indicated by a sinusoid stroke and a diagonal dash to its right: ʃ', except for one instance where it has the L-shape (p. 1 last line).

The symbol for 2/3 on p. 3 line 23 looks like sampi for 900: ϡ. However, on p. 1 line 20 it may be possible to construe it as a cursively written beta with curved stroke to the right: υ)[2].

Ordinals appear in one problem only and are either written out as words or indicated ὁ ᾱ, ὁ β̄.

The following abbreviations and symbols are used:

$\alpha\pi\overset{\lambda}{\eta}$ = ἀπηλ(ιώτου)	$\pi\overset{\chi}{\eta}$ = πήχ(εις)	ϛ = ἅ(μματα)	< = (ἀριθμός?)
λιβ' = λιβ(ός)	γt = γί(νονται)	λοιπ' = λοιπ(όν)	* = (δηνάριον)
νοτ' = νότ(ου)	S' } (ἥμισυ)	εχ' = ἔχ(ει)	ΓΊ = (μυριάς)
βορ' = βορ(ρᾶ)	L } (ἥμισυ)		ϡ = 2/3

[1] This is not an exhaustive listing. Errors are noted in the line-by-line commentary.

[2] Numerous methods of expressing 2/3 are known: see the lists in *MPER* n.S. 1, p. 50; Brashear, *ZPE* 60 (1985) 242.

περιφερ/ = -φέρεια	ᾱ = (πρῶτος)	τω̄ = τῶ(ν)	ο̄ = (ἀρτάβη)
υφελ/ = ὕφελ(ε)	β̄ = (δεύτερος)	κρατησο̄ = κράτησον	λπται = λ(εί)π(ε)ται
ποσω̄ = πόσω(ν)	παρ/ = παρ(ά)	μοριο̄ = μόριο(ν)	αρουρ/ = ἄρουρ(α, - ας etc.)
πηχ/ = πήχ(εις)	ομο⁺ = ὁμοί(ως)	πρ$\overset{κ}{\bar{ο}}$ = προκ(είμενος)	αρουρω̄ = ἀρουρῶ(ν)
		ζ = (τάλαντον)	

The mathematical vocabulary is comparable to those in *P. Achmim*, pp. 13ff.; *MPER* n.S. 1, pp. 41-6; and *P. Mich.* III, pp. 46-7:

σύνθες (ὁμοῦ) add (together)
σύνθες ἐπί multiply
μέρισον divide
ἐπί multiply
παρά divide
ὕφελε substract
δίπλωσον double
τετράπλωσον quadruple
τρίπλωσον triple
πολυπλάσιασον multiply
... καὶ ... add
ἐφ' ἑαυτά to square
λοιπόν remainder
λείπεται remainder
χωρέω contain
ψῆφος number?

περιφέρεια circumference
ἐμβαδόν cubic
νέμησις area
μόριον fraction
μῆκος length
μλάτος breadth
βάθος depth
ὕψος height
λίθος stone = rectangular solid
θησαυρὸς στρογγυλοῦν round granary = truncated cone
ναύβιον τετραγωνοειδές rectangular trench
measures: ναύβιον
ἄρουρα
πῆχυς
ἅμμα
κάλαμος
ἀρτάβη
σχοίνιον

In addition to the aforementioned orthographical and syntactical errors there are also occasional factual errors, e.g. σφραγίς instead of ἄρουρα (p. 3 1.12); wrong units of measure (p. 2 1.23); an incorrectly labelled drawing (p. 1). Furthermore, computational errors (p. 1 lines 21, 23; p. 3 lines 11-12) mar the quality of the work. One possible source for all of these errors might have been a faulty *Vorlage* from which the person writing these exercises either copied directly or else had read to him while he wrote from dictation. (The original source of error in the putative faulty *Vorlarge* was probably simple miscalculation: whoever was originally responsible for creating and computing these problems erred, thus leading astray perhaps whole generations of students and teachers after him who blindly copied his erroneous calculations without ever bothering to check for themselves their correctness). Another possible explanation for the errors might be that this codex is the *Urtext* itself — there was no *Vorlage*. The mathematician writing these problems made the mistakes himself while computing them. In the case of other mistakes, one gets the impression that the copyist misunderstood the (presumably correct!) text as it was dictated to him.

The following examples will serve to illustrate the theories proposed above:

The same error remarkably appears in two different problems: ϛ̓ (6000) is mistakenly written for Ἀ (1000) both on p. 3 line 4 and p. 4 line 2. It is difficult to imagine anyone misunderstanding either verbally or visually ἑξαχίλιος for χίλιος or ϛ̓ for Ἀ. In this case one could presuppose either a faulty *Vorlage* or a mistake made in computation by the person while he worked out the problem himself.

At least one error — the missing line on p. 2 after line 16 — is most likely due to a mistake committed during copying or dictating. The writer (or the person dictating), seeing the same wording at the beginning of both lines, missed the first line and caught only the second line - provided, of course, that the text being copied or dictated from was correct and complete. Otherwise, a faulty *Vorlage* could have been the source of error.

On the other hand, εφροχου on p. 2 line 7 is most easily explained as a misspelling of a poorly dictated ἀβρόχου or ἐμβρόχου. The same holds true for the errors of vowel quantity committed throughout these exercises: τόν for τῶν or πόσον ναύβιον for πόσων ναυβίων. Yet from all we know about Greek and Roman pedagogical methods it is only the literary exercises that were dictated (S. Bonner, *Education in Ancient Rome*, London 1977, 177, 180; L. Grasberger, *Erziehung und Unterricht im klassischen Altertum* II, Würzburg 1875, repr. Aalen 1971, 298ff.). Elementary word exercises and banal addition, multiplication and division tables were certainly not dictated (*pace* Boyaval, *Rev. arch.* 1973, 24; Brashear, *REG* 97 (1984) 217, 4). Whether longer mathematical problems such as these were copied or dictated is not something ancient authors thought necessary to mention. Hence further speculation on the topic is pointless.

Sometimes some of the steps in the calculations are omitted (this phenomenon is also characteristic of the problems in *P. Achmim*). Furthermore, the actual computations were done on fingers, wax tablets, ostraca or scrap, examples of which may be *P. Petrie* III, pp. 98-99; *P. Oxy.* 2550; G. Plaumann, *Amtl. Berichte aus den königl. Kunstsammlungen* 34, Berlin 1913, 223 (= Pack[2] 2739). For clarity's sake I have occasionally made explanatory insertions in the Greek text in angular brackets < > and discussed the procedures in the introductions to the individual problems.

At other times, the subtotals and intermediate calculations are repeated or signalled by τοῦτ' ἔστιν, or otherwise unnecessarily — and redundantly — written out.

The problems are all examples of applied mathematics and geometry and relate directly to daily life — how to calculate the volumes of a stone, trench and a granary, the areas of fields, proportionate sums of money in talents and denarii. The methods used to solve the problems are prosaic and goal-oriented — little time or space is wasted on explaining the whys and wherefores of any of the procedures involved. This pragmatic approach, characteristic of the majority of the mathematical texts from Egypt, is shared with the Heronian corpus with its "auf den praktischen Gebrauch zugeschnittene Behandlung des Stoffes mit nur wenigen Hinweisen auf die verwendete Methode" (Vogel, *MPER* n.S. 1, p. 35; Mau, *APF* 17 [1960] 10). Only rarely does the occasional question interrupt the flow of the solution, and then it usually has nothing to do with theory but again only with practical matters: what is the numbers of cubits in a schoenion? the number of cubic cubits in an artab? etc.[3]

While it can be neither proven nor refuted that these problems were a product of Pachomian erudition and diligence, considering their placement in the codex, the hypothesis suggested by J. Robinson seems the most attractive: the mathematical problems were probably already in the codex if the monks bought it ready-made or already on the sheets of papyrus used by the monks to make the codex[4]. Instead of deleting or excising the problems they simply wrote around them. (The opposite

[3] Similar in style are also Didymus (ed. Heiberg, Copenhagen 1927); the geometrical exercises published by J. Rudhardt, *Museum Helveticum* 35 (1978) 233f.

[4] E. Turner, *The Typology of the Codex*, Univ. of Pennsylvania 1977, 73f., discusses the alternative possibilities. B. van Regemorter, "Le papetier-libraire en Egypte", *Chronique d'Egypte* 35 (1960) 278-280, notes examples of "re-cycled" papyrus leaves used for making codices.

I would like to thank A. Wouters (Louvain) for checking some of the readings on the originals and D. Fowler (Coventry) for improving the mathematical commentary.

situation is the case with the Coptic calculation manual published in *BSAC* 13 [1948-9] 151 ff. There the earlier Greek-Coptic Bible text was expunged to make way for the mathematical exercises).

O. Guéraud and P. Jouguet, *Un livre d'écolier*, Cairo 1938, xiv conclude that the school exercises on their papyrus roll were not written by a student but by someone for the student — "c'est une sorte de manuel, où l'enfant pouvait s'exercer à lire et à compter". H.-I. Marrou, Geschichte der Erziehung im klassischen Altertum, Freiburg-Munich 1957, 227, concurs. Discussing bilingual word-lists in *Proceedings of the XIVth International Congress of Papyrology*, Chico 1981, 31-41, I stressed the same point (cf. J. Kramer, *Glossaria Bilinguia*, (PTA 30), Bonn 1983, 10; id., *Atti del XVII Congresso Internazionale di Papirologia* III, Naples 1984, 1377 f.). Were the problems in this codex written by a student? Or were they part of an instruction manual prepared by the teacher (from another copy? composed ad *hoc*?)? for his own consultation? or for the aid of the student? In view of the facile, expert writing and the nature of the mathematics involved, these exercises were evidently written by someone advanced both in the art of writing and reckoning (cf. G. Plaumann, *Amtl. Berichte aus den königlichen Kunstsammlungen* 34, Berlin 1913, p. 223) — whether pupil or teacher is just one of the many imponderabilia confronting us with regard to this codex.

On the crude little sketches which sometimes accompany the geometrical problems see K. Weitzmann, Illustrations in Roll and Codex, Princeton 1970, 40 ff.

Recently published parallels are in R. Pintaudi, P.J. Sijpesteijn, Tavolette ligne e cerate da varie collezioni (P. Flor. XVII), Florence 1989.

TRANSCRIPTION, TRANSLATION, NOTES

Extant page one

↓ …[ ]…

ναύβιον τετραγωνοε̣ι̣δ̣[(ὲς) νότ(ου)] πηχῶν λη βο[ρ(ρᾶ)]
$\overline{\lambda\varsigma}$ ἀπηλ(ιώτου) $\overline{\kappa}\eta$ λιβ(ὸς) $\overline{\kappa}\varsigma$ βά̣θ̣[(ος)] πηχ(ῶν) $\overline{\delta}$. πόσο̣ν να[ύ-]
βιον; ποίει οὕτω· σύνθες νότου καὶ βορ(ρᾶ) … —
τοῦτ᾽ ἔστιν [λ]$\overline{\eta}$ καὶ $\overline{\lambda}\varsigma$ γί(ν.) $\overline{o}\delta$ ὧν τὸ ʃ´ γί(ν.) $\overline{\lambda}\zeta$. ὁμο[ί(ως)]
[σύν]θες ἀπηλιώτου καὶ λιβός, τοῦτ᾽ ἔστι̣ν
[κη] καὶ $\overline{\kappa}\varsigma$, γί(ν.) $\overline{\nu}\delta$ ὧν τὸ $\overline{ʃ}$´ γί(ν.) $\overline{\kappa}\zeta$. πολυπλασίασ[ον]
[{πολ]υπλασίασον} $\overline{\kappa}\zeta$ καὶ $\overline{\lambda}\zeta$ [ἐπ]ὶ̣ τὸ βάθος
πηχῶν $\overline{\delta}$, ταῦτα [μέρισον παρ(ὰ) $\overline{\kappa}\zeta$ γί(ν.) $\overline{\rho\mu\eta}$]
διὰ τί παρὰ τὸν $\overline{\kappa}\zeta$; διότι τὸ ναύβιο[ν ἔχει]
— πήχ(εις) $\overline{\kappa}\zeta$.[
ἀπηλ(ιώτου) $\overline{\kappa}\eta$ αLd αι/

νότ(ου) $\overline{\lambda\eta}$ | βάθ(ος) δ | βο̣ρ̣(ρᾶ) $\overline{\lambda\varsigma}$

ξ

λιβ(ὸς) $\overline{\kappa\varsigma}$

2-14 Calculating the volume of an irregular rectangular trench in cubits and naubia. The nominal volume in cubits is obtained by calculating the average length and width, then multiplying these averages by the depth. The result is the volume in cubits. To find the volume in naubia one divides the cubic cubits by 27, since 1 naubion = 27 cubic cubits.
2 ναύβιον: The word has three meanings: 1) a trench — as here; 2) a cubic measure — as in line 9; 3) a tax — see S. Wallace, *Taxation in Egypt from Augustus to Diocletian*, Princeton 1938. There seem to be traces of writing in the upper margin, but neither A. Wouters nor I can discern anything legible.
3-4 l. πόσω̣ν̣ ναυβίων.
4 βορ´ λ̣η̣κ̣̄: Perhaps numbers which the writer had already penned before opting for the τοῦτ᾽ ἔστιν construction.

2-14 "(Given) a rectangular tranch of 38 cubits (to the south), 36 to the north, 28 to the east, 26 to the west (and) 4 cubits deep, how many naubia is it? Do it this way. Add (the) south and north, ... that is 38 and 36, equals 74, 2' of which is 37. Likewise, add (the) east and west, that is (28) and 26, equals 54, 2' of which equals 27. Multiply 27 by 37 by the depth of 4 cubits. (Divide) this (by 27. The result is 148). Why (divide) by 27? Because the naubion (contains) 27 (square) cubits.

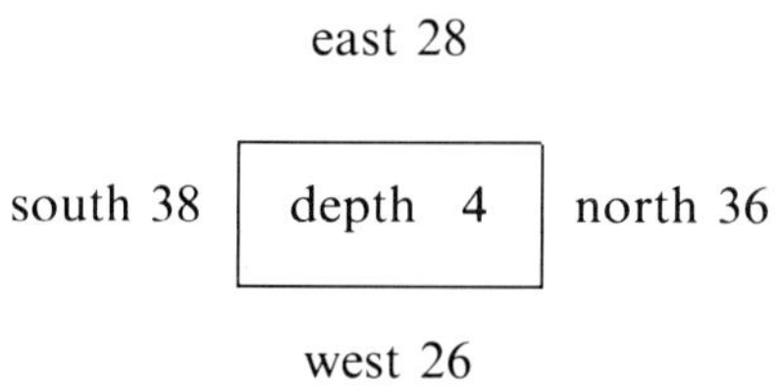

5,6 Pap. τουτ̓.
10 παρὰ τῶν.
PSI 186.6f. seems to be the only other papyrus containing such a question as those found, *passim*, here in these problems. J. Shelton, *ZPE* 12 (1981) 100, cites as parallel Heron, *Stereometrica* 7.
10-11 τὸ ναύβιο[ν ἔχει] πήχ(εις) κ̅ζ̅: This conforms to the definition given in *P. Oxy*. 2847.24-26: ναύβιον δέ ἐστιν μέτρον [γ]ῆς ὀρυσσόμενον τρεῖς πήχεις ἔχ[ο]ν τοῦ πλά[του]ς καὶ μήκους καὶ βάθους. Cf. *P. Oxy*. 1053 introd., 669.11; H.C. Youtie, *TAPA* 72 (1941) 443 n. 15. The drawing is neither to scale nor is it correctly labelled. To its right is another diagram which must refer to another problem now lost, since the numbers have nothing to do with the calculations on this page. (Visible above and to the right is a curved stroke of ink).

ϑησαυρὸς στρ[ο]γ<γ>υλοῦν· κάτω περιφέρ(εια)
πηχ(ῶν) κ̅δ̅ ἄνω περιφέρ(εια) πηχ(ῶν) ⟦κδ⟧ ι̅ϛ̅ τὸ δὲ
ὕψος πηχ(ῶν) ε̅· εὑρεῖν πόσας (ἀρτάβας) χωρήσει
ὁ ϑησαυρός· σύνϑες κ̅δ̅ καὶ ι̅ϛ̅ γί(ν.) μ̅
ὧν τὸ 𐅵 γί(ν.) κ̅ ἐφ' ἑαυτὰ γί(ν.) υ. τοῦτων τὸ ι̣̅β̅
γί(ν.) λγγ''. ταῦτα ἐπὶ τὸ ὕψος πηχ(ῶν) ε̅ γί(ν.) ρ̅ξ̅ϛ̅β̅'
ταῦτα ἐπὶ γ̅d̅η̅, ρ̅ξ̅ϛ̅β̅' οὖν ἐπὶ γ̅d̅η̅, γί(ν.) φ̅ι̅β̅𐅵
ὥστε ὁ ϑησαυρὸς οὕτως χωρήσει
(ἀρτάβας) φ̅ι̅β̅𐅵
ἀπὸ α̅ ὕφελ(ε) γ'' ι̅ε̅. ἐν ποίῳ ψήφῳ
ἔρχεται γ'' ι̅ε̅; τῶ(ν) β̅ τὸ ε̅ γ'' ι̅ε̅.
ἀπὸ τῶν ε{''} ὕφελ(ε) β̅, λ(εί)π(ε)ται γ̅· τῶν γ̅ τὸ ε
Lι''

15-23 Calculating the nominal volume in artabs of a granary having the shape of a truncated cone, given the upper and lower circumferences of 16 and 24 cubits respectively and the height of 5 cubits. The formula in use for the nominal volume was $\frac{h}{4\pi}C_m^2$ where C_m is the mean circumference. Inserting the values of this problem results in

$$\frac{5}{4\times 3}\left(\frac{24+16}{2}\right)^2 = 166\ 2/3 \text{ cubic cubits}$$

To find the capacity in artabs this number is multiplied by 3 3/8, the number of artabs in a cubic cubit:

166 2/3 × 3 3/8 = 562 1/2 artabs (not 512 1/2 as appears in lines 21 and 23!).

Parallels are analyzed and discussed by J. Shelton, *ZPE* 42 (1981) 99ff.; B. Boyaval, *ZPE* 28 (1978) 203-5; O. Neugebauer, *Vorgriechische Mathematik*, Berlin 1934, 126; R. Parker, *Demotic Mathematical Papyri*, Providence-London 1972, 54.

15 Instead of the stop after στρογγυλοῦν one can emend οὗ ἡ κάτω π. ... καὶ ἡ ἄνω π. Both usages are attested.

21,23 l. φξβ𐅵

15-23 "There is a circular granary. A lower circumference is 24 cubits, an upper circumference 16 cubits and the height 5 cubits. Find out how many artabs the granary will hold. Add 24 and 16, equals 40, 2' of which is 20. This times itself equals 400. 12' of this equals 33 3'. This times the height of 5 cubits equals 166 3''. This times 3 4'8', (that is) 166 3'' times 3 4'8', equals 512 2' (sic! read: 562 2'), so that the granary will thus contain 512 2' (sic! read 562 2') artabs".

24-27 "From 1 subtract 3' 15'. In which cipher does 3' 15' go? 5' of 2 (is) 3' 15'. From 5 subtract 2, leaves 3. 5' of 3 (is) 2' 10'."

21 The rendition of η′ = 8' varies considerably from one instance to the next, resembling sometimes a y sometimes an s and indistinguishable from the symbol for 2'. Cf. the cursive form used in the fraction 38' on pp. 3,4 of the codex.
The equation 1 cubic cubit = 3 4'8' artabs is a standard one. J. Shelton, *op. cit.*, 100 cites parallels.
22 l. οὗτος?
24-27 Substraction problem: 1-3' 15' = 2' 10'. The procedure is explained by J. Baillet in *P. Achmim*, p. 48. Instead of fractions whole numbers are substituted that have the same proportional relationship to one another as the fractions, and the subtraction is computed with them. When the subtraction has been completed, the resulting whole numbers are converted back into fractions. Applying this principle to the problem at hand:

3' 15' = 2:5
1 × 5 = 5
5-2 = 3
3:5 = 2' 10'

Cf. R. Parker, *Demotic Mathematical Papyri*, Providence-London 1972, 67-8.
24 l. ποίᾳ
26 The two diagonal strokes after epsilon should be deleted; otherwise they make epsilon into the fraction 5'.
Pap. λπται: Cf. *P. Achmim*, *passim*: λπται.
27 The symbol for 2' here (L) is different from the sinusoid curve (ʃ′) used elsewhere in these problems.

Extant page two

→ [λί]θος οὗτος μὲν [μῆ]κος πηχ(ῶν) β̅, τὸ πλάτος
πηχ(ῶν) β̅, βάθ(ος) πηχ(ῶν) γ̅. εἰδέναι πόσω(ν) ⟦ὁ λίθος οὗτος⟧
[ἔστ]ạι ὁ λίθος ἐμβαδῶν πηχ(ῶν) ε̅. σύνθες τὸ μῆκος
ẹπί τὸ πλάτος, τοῦτ' ἔστιν β̅ ἐπὶ β, γί(ν.) δ ẹ[πὶ τὸ]
βạ́θος πηχῶν γ̅ γί(ν.) ι̅β̅, ὥστε ὁ λίθο[ς ἔχει]
πήχ(εις) ι̅β̅
[ ]ωνεξεφροχου ἀρουρ[( )]ρ̣[.]
[ ]. .[. . .]. .οῦντες β̅, ὁ (πρῶτος) ἔχει ἀρού-
ρ̣ạ[ς] τ̅, ọ (δεύτερος) Σ̅. οὕτω ποίει· ρ̅ ἐπὶ τοῦ (πρώτου)
γί(ν.) (μυριάδες) γ̅. ταῦτα μέρ⟦ọ⟧ισον παρ(ὰ) φ̅ γί(ν.) ξ̅, ὥστε ἀπὸ
τοῦ πρώτου ἀρουρῶν ξ. πάλιν ὁμοί(ως) ρ̅ ἐπὶ τοῦ
(δευτέρου) γί(ν.) (μυριάδες) β̅. ταῦτα μέρισον παρ(ὰ) φ̅ γί(ν.) μ̅, ὥστε
ἀπὸ τοῦ (δευτέρου) ἀρουρῶν μ̅. σύνθες μ̅ καὶ ξ̅ <γί(ν.) ρ̅>.
σφραγὶς νότου σχοινία ι̅β̅ βορ(ρᾶ) σχοινία ι̅β̅
ἀπηλ(ιώτου) πηχῶν η̅ λιβ(ὸς) πηχ(ῶν) η̅. πόσων

1-6 Calculating the volume of a stone measuring 3 × 2 × 2 cubits. Parallels can be found in *MPER* n.S. 1, p. 16f.

3 ἐμβαδῶν πηχ(ῶν): Cf. *P. Oxy.* 921 vso: πήχεις ἐμβαδοί and Heron, *De mensuris* (ed. Heiberg) 23, where the same adjectival form appears.

The substantive ἐμβαδόν generally indicates surface area, and *P. Oxy.* 669 distinguishes very clearly between linear (εὐθυμετρικός), square (ἐμβαδικός) and cubic (στερεός) cubits. Vogel, *MPER* n.S, 1, p. 42.1, however, cites instances where ἐμβαδόν also means volume. Conceivably the writer intends that meaning here. The significance of the number 5 following it is obscure. It presumably represents the sum of the depth and one of the side-lengths, but it is irrelevant to the problem at hand.

4 Pap. τουτ̓.

5-6 λίθο[ς ἔσται] πηχ(ῶν) is also possible.

7-13 Although the first two lines are damaged beyond recovery the point of the problem is not difficult to deduce. Two fields measure 300 and 200 aruras respectively. To 100 aruras of the total area of 500 aruras something has happened. Depending on how one emends the curiously misspelled word in l. 7, the 100 aruras have been either left unwatered by the Nile's waters (reading ἄβροχος) or else they have been entirely inundated and are still lying under water and are hence uncultivable (reading ἔμβροχος). The problem is to find the proportionate amounts in both fields which added together make up the 100 aruras.

1-6 "There is such a stone: 2 cubits long, 2 cubits broad, 3 cubits deep. Find out how many cubic (?) cubits such a stone will be. Multiply the length by the breadth, that is 2 times 2 equals 4, (times the) depth of 3 cubits, equals 12, so that the stone (has) 12 (cubic) cubits".

7-13 "... 100 aruras of which are unwatered/flooded (?) ... 2 fields lying fallow (?), the first having 300 aruras, the second 200. Do it this way. 100 times the first (field) equals 30,000. Divide this by 500, equals 60, so that from the first there are 60 aruras. Again, likewise, 100 times the second (field) equals 20,000. Divide this by 500, equals 40, so that from the second there are 40 aruras. Add 40 and 60, (equals 100)".

14-21 "There is a field of 12 schoenia on the south, 12 schoenia on the north, 8 cubits on the east, 8 cubits on the west. How many aruras is the field? Add 12 (and) 12 (2' of which is 12. Add 8 and 8), 2' of which is 8. Likewise, 8 times 12 equals 96. The schoenion has 96 cubits. Divide, therefore, 96 by 96 equals 1, so that the field has 1 arura".

The problem is solved in much the same way as it would be today:

$$300+200 = 500. \qquad \frac{100 \times 300}{500} = X \qquad \frac{100 \times 200}{500} = Y$$

$$X+Y = 100$$

7 1. ἀβρόχου? ἐμβρόχου? The long vertical stroke at the end of the line presumably belongs to ρ = 100.
8 κλ]ῆ̣ρ̣[ο̣ι ἀ]ρ̣γ̣οῦντες?
10,12 ṅ γ, ṅ β̄: Various forms of this symbol are discussed in *ZPE* 60 (1985) 239ff.
13 μ̄ καὶ ξ̄ has been written over something or else been re-written.
14-21 Calculating the area of a rectangular field measuring 12 schoenia by 8 cubits in aruras. Lengths of parallel opposite sides are added: 12+12 = 24 and 8+8 = 16, and the averages taken: 12 and 8. These numbers multiplied (96) and divided by the number of cubits in a schoenion (96) result in the number of square schoenia in the field (1). Since one square schoenion is equal to one arura, the answer is one arura. According to D. Crawford, *Kerkeosiris*, Cambridge 1971, 12, owing to the nature of the irrigation system in Egypt, long, narrow fields, such as the one described here, predominated.
14 1. σχοινίων *bis*.

ἀρουρῶν \ἡ/ σφραγίς; σύνθες $\overline{\text{ιβ}}$ $\overline{\text{ιβ}}$
<ὧν τὸ ʃ´ γί(ν.) $\overline{\text{ιβ}}$. σύνθες $\overline{\text{η}}$ καὶ $\overline{\text{η}}$>
ὧν τὸ ʃ´ γί(ν.) η. ὁμοί(ως) η ἐπὶ $\overline{\text{ιβ}}$ γι(ν.) $\overline{\text{ϙς}}$.
τὸ σχοινίον ἔχει πήχ(εις) $\overline{\text{ϙς}}$. $\overline{\text{ϙς}}$ οὖν
{ταῦτα} μέρισον παρ(ὰ) $\overline{\text{ϙς}}$ γί(ν.) $\overline{\text{α}}$, ὥστε
ἡ σφραγὶς ἀρουρῶ(ν) $\overline{\text{α}}$
ἄλλην μέθοδον ποιοῦμαι· οὕτω ποίει·
ϙς· πόσα ἅμ<μ>ατα ἔχει; ἔχει $\overline{\text{ιβ}}$. $\overline{\text{ϙς}}$ οὖν
παρ(ὰ) $\overline{\text{ιβ}}$ γί(ν.) $\overline{\text{η}}$. η ἐπὶ η γί(ν.) ξδ. ταῦτα
μέρισον παρὰ τὸν ξδ γί(ν.) $\overline{\text{α}}$· διὰ τί
παρὰ τὸν $\overline{\text{ξδ}}$; διότι ἡ ἄρουρα ἔχει
ἅμ(ματα) $\overline{\text{ξδ}}$

17 Here a whole line has evidently dropped out through haplography.
20 The writer assumes the equation 1 schoenia^2 = 1 arura is known.
21 l. ἀρούρας.
22-27 Calculating the area of the same field in hammata. Starting from the number 96 (taken from l. 17 in the preceding section) the writer computes the number of hammata in 96 cubits. Since one hamma contains 12 cubits, the number of hammata in 96 cubits is eight. The field in question therefore has 64 square hammata or 1 square arura, since 1 arura = 64 square hammata. A similar problem is found in *ZPE* 42 (1981) 12.

22-27 "I do another method. Do it this way. 96. How many hammata does it have? (sic! Read: 'How many cubits does the hamma have?). (The hamma) has 12 (cubits). Therefore, (divide) 96 by 12, equals 8. 8 times 8 equals 64. Divide this by 64, equals 1. Why by 64? Because the arura has 64 (square) hammata".

23 l. πόσας πήχεις ἔχει τὸ ἅμμα; τὸ ἅμμα ἔχει πήχεις $\overline{\iota\beta}$.
ἅμμα: J. Shelton, *ZPE* 42 (1981) 94.2, defends the *spiritus asper* as opposed to other modern editors who give it the *lenis*.
25,26 l. παρὰ τῶν.
27 This symbol/abbreviation seems to consist of a very cursively written *Hakenalpha* (⅃) which merges into either a vestigial mu (∩) or an abbreviation stroke, depending on how one wants to interpret it. The only other abbreviation for ἅμμα I could find is in *P. Berl. Leihg.* II 31 passim, where it consists of alpha topped by a horizontal bar (G. Poethke, *per litt.* 8. April 1985).

Extant page three

↓ σ̣φραγὶς νότου καλά[μων λ]ε̅· τί δῶ μὲν τ̣[. . .] ρ̣ι οἷα νέμησις εἰς τὸ . . . . [. . ε]ὑ̣ρεθῆναι ⟦κ⟧β̅[Ld?] οὕτω ποίει· ἡ μι<ὰ> ἄρουρα ἔχει κάλαμοι υ̅· υ̅ οὖν̣ ἐπὶ β̅L̅d̅ γί(ν.) ϛ̓̅ρ̅. τί ἐπὶ τί ἵνα γένηται ϛ̓̅ρ̅; μ̅ε̅ π:[.] κ̅δ̅ γί(ν.) ϛ̓̅ρ̅. μ̅ε̅ δίπλωσον γί(ν.) q̅. ὕφ̣[ελ](ε) λ̅ε̅ λοιπ(ὸν) ν̅ε̅ ὥστε βόρ(ρᾶ) ν̅ε̅ κ̣[αλάμων? . . . .] ἀπηλ(ιώτου) κ̅δ̅ λιβ(ὸς) ἴσος. σύνθ[ες νότου] καὶ βορ(ρᾶ) λ̅ε̅ καὶ ν̅ε̅ γί(ν.)q ὧν τὸ 𐅵 γί(ν.) μ̅ε̅ ὁμοί(ως) σύνθες ἀπηλ(ιώτου) καὶ λιβ(ός), κ̅δ̅ καὶ κ̅δ̅, ὧν τὸ 𐅵 γί(ν.) κ̅δ̅. πολυπλασίασον οὖν μ̅ε̅ ἐπὶ κ̅δ̅ γι(ν.) ϛ̓̅ρ̅· ταῦτα μέρισον παρ(ὰ) υ̅ γί(ν.) β̅L̅d̅. διὰ τί παρ(ὰ) υ̅; <δι>ότι ἡ σφραγὶς ἔχ(ει) κάλαμοι υ̅

1-12 The problem seems to be the following: given a field whose southern side measures 35 kalamoi, compute the boundaries of the field such that it measures 2 2' 4' aruras. The essential information that one arura = 400 (square) kalamoi is immediately volunteered. The field in question would then measure 1100 square kalamoi (not 6100 as given in the text).

The nominal area is calculated by multiplying the averages of the two pairs of opposite sides:

$$1100 = \tfrac{1}{2}(a+35) \times \tfrac{1}{2}(b+d)$$

As possible average side measurements for the field 45 × 24 = 1100 are suggested (actually 45 × 24 = 1080, neither 6100 as given in the text, nor 1100). Since the ancient Egyptians, and the Greeks, Copts and Arabs after them, computed the area of fields using the averages of opposite sides, the writer here doubles 45. 90 then is the theoretical double of two opposite sides. Since one factor (the south side measuring 35 kalamoi) is known, we can substract it from the theoretical double to obtain the length of the north side: 90 − 35 = 55 kalamoi on the north side. Somehow it is determined that the east and west sides are equal and consist of 24 kalamoi each. These values for the four sides are then checked:

$$\tfrac{1}{2}(35+55) \times \tfrac{1}{2}(24+24) = 1080 \text{ (neither 1100, nor 6100!)}$$

1-12 "(Given) a field of 35 kalamoi to the south, what should I add ... such an area as to be found (measuring) 2 (2' 4' aruras?)? Do it thus. One arura has 400 kalamoi. 400, therefore, times 2 2' 4' is 1100 (not 6100!). What times what equals 6100 (sic!)? 45 (multiplied by) 24 equals 6100 (sic!) Double 45, equals 90. Subtract 35 (from 90) leaving 55, so that the north is 55 (kalamoi) ... the east 24, the west is equal. Add (south) and north, 35 and 55 is 90, 2' of which is 45. Likewise, add east and west, 24 and 24, 2' of which is 24. Multiply, therefore, 45 times 24 is 6100 (sic!) Divide this by 400 equals 2 2' 4'. Why by 400? Because the field (sic! read: the arura) contains 400 kalamoi".

The area in square kalamoi is checked by diving by 400 which is the number of square kalamoi per arura: $\frac{1100}{400} = 2\ 3/4$ aruras

For similar problems see J. Shelton, *ZPE* 42 (1981) 92ff.

1 δῶ μεν? δῶμεν?

3 Was μι' ἄρουρα intended?

4 l. 'Aς *bis*, instead of ϛ̓ρ. The second ϛ̓ has been written over something. The same error - ϛ̓ for 'A - appears on p. 4 line 2.
πọ\λ̣/[(υπλασιάζεται ἐπί)]?

5 δίπλωσον: According to H. Cuvigny and G. Wagner, *ZPE* 62 (1986) 69, this verb was heretofore unattested in Egypt, making its first appearance in a letter they publish there.

10 l. 'Aπ.

12 l. ἡ ἄρουρα ἔχει καλάμους $\bar{υ}$.

σφραγὶς νότου κάλαμοι ο̅ βορ(ρᾶ) κάλαμοι ν̅
ἀπηλ(ιώτου) κάλαμοι λ̅ε̅ λιβ(ὸς) κάλαμοι μ̅ε̅. πόσω(ν)
ἀρουρῶν ἡ σφραγίς; σύνθες νότ̣ο̣υ̣
καὶ βορρᾶ ὧν τὸ ʃ´, ὁμοί(ως) ἀπηλ(ιώτου) καὶ λιβὸς
ὧν τὸ ʃ´, ἐφ᾽ ἑαυτὰ γί(ν.) ͵Β̅υ̅. ταῦτα μέρισο(ν)
παρ(ὰ) υ̅ <γί(ν.)> ς̅, ὥστε ἀρουρ(ῶν) ς̅.

13-14 l. καλάμων *quater*.
13-18 Calculating the area of an irregular quadrilateral field measuring 35 × 50 × 45 × 70 kalamoi in aruras. After obtaining the area in square kalamoi (2400) we divide it by 400, the number of square kalamoi per arura, and get the correct answer of 6 square aruras. The procedure for calculating the area is only approximate, since $\frac{1}{2}(a+c) \times \frac{1}{2}(b+d) = \frac{1}{4}(ab+bc+cd+da)$, while the actual area is $\frac{1}{4}(ab \sin ab + bc \sin bc + cd \sin cd + da \sin da)$. If any vertex is not a right angle (so that the field is not rectangular), then the corresponding term, such as ab sin ab, will be diminished to ab. Hence, for non-rectangular fields, the procedure will always over-estimate the area.
Already attested in acient Egypt, this method continued to be used there down to the beginning of this century. For parallels and discussion see T.E. Peet, *The Rhind Mathematical Papyrus*, Liverpool 1923, 94; F.G. Kenyon, *P. Lond. II*, p. 129ff.; B. Grenfell, *P. Tebt. I*, p. 385ff.; J. Shelton, *ZPE* 42 (1981) 96-7; D. Crawford, *Kerkeosiris*, Cambridge 1971, 12; O. Montevecchi, *Aegyptus* 63 (1983) 12-14; K. Vogel, *Die vorgriechische Mathematik*, Hannover-Paderborn 1958, 59, 66; Heron, *Geometrica* 6.2 (ed. Heiberg); R. Parker, *Demotic Mathematical Papyri*, Providence-London 1972, 71.
Another method would involve dividing the field into several rectangles and triangles whose area could be calculated exactly. The sums of these areas would give exact total area. Cf. Boyaval, *ZPE* 28 (1978) 210; *P. Chic. Lit.* 3, p. 24.
13-14 The sequence south, north, east, west is also hallowed by age, cf. G. Posener, "Sur l'orientation et l'ordre des points cardinaux chez les Egyptiens", *Nachrichten d. Akad. d. Wissens. Göttingen, Phil.-Hist. Kl.* 1965, pp. 69-78; Brashear, *BGU* XIV 2398.9 n.; and P.D.A. Harvey, *Topographical Maps*, London 1980, index *s.v.* orientation, for later traditions.
17 Pap. ͵Β = 2000.
18 ὥστε ἀρουρ(ῶν) ς̅ is so cursive and scrawly that it could be by a different hand.
The equation 1 arura = 400 square kalamoi occurs three times in the Chester Beatty mathematical texts — twice verbatim in lines 3 and 12 on this page (instead of σφραγίς read there **ἄρουρα**) and a third time implicit in the division here. A square-shaped arura of this area would have sides measuring 20 kalamoi each. As is the case with most of the measures whose amounts differed over the course of the centuries and according to the localities, one finds

13-18 "(Given) a field of 70 kalamoi on the south, 50 kalamoi on the north, 35 kalamoi on the east, 45 kalamoi on the west, how many aruras is the field? Add south and north; (calculate) the 2'. Likewise east and west; (calculate) the 2'. (The halves multiplied) by each other give 2400. Divide this by 400, (equals) 6, so that it is 6 aruras".

kalamoi calculated at anywhere from 4 to 6 2/3 cubits. Heron, *Geometrica* 4.11, works with a kalamos of 5 cubits. Taking this kalamos of 5 cubits and our theoretical square of 20×20 kalamoi would result in an arura of 100×100 cubits. Just such an arura is attested from Egypt — v.U. Wilcken, *Grundzüge der Papyruskunde*, Leipzig 1912, lxxii.

19-23 Calculating with parts: If 3' 14' 42' of an unknown number (n) of talents is 200, what is the 42nd part of n? 3' 14' 42' is first identified as coming from a table of 7ths, viz. the 7th of 3, although this fact is subsequently never used in the computations. 3' 14' 42' of 42 gives $(14+3+1=)$ 18. Hence, the scribe seems to reason: if 3' 14' 42' of n = 200 and if 3' 14' 42' of 42 = 18 then 42' of the unknown n is 18' of 200 or 11 9'. Thus n = 42×11 9' = 466 3" (the scribe mistakenly writing/calculating 666 3").

20 1. ἐν ποίᾳ ψήφῳ: The same question appears in *P. Achmim passim*, where Baillet translates it "dans quel calcul?" and K. Vogel, *Sitzungsber. d. bayer. Akad. d. Wiss.* Math.-Naturhist. Abt. 1936, pp. 404, 429; "in welcher Tabelle (kommt die zu teilende Zahl vor)?". Both evidently regard ψῆφος as a division table, such as one finds at the beginning of *P. Achmim*, which the person working the problem would consult to find the answer to this question. Furthermore, the phrase ἐν ψήφων has recently appeared as part of the heading in a number of division tables: P. Sijpesteijn, *CdE* 56 (1981) 101; Wm. Brashear, *Enchoria* 12 (1984) 3, id., 13 (1985) 21f., id., 14 (1986) 2f., id., *BSAC* 26 (1984) 19ff., id., *REG* 97 (1984) 214-216; P. Cauderlier, *Rev. archéol.* 1983, 259f. Various attemps have been made to elucidate it, none of them ultimately satisfactory. In *Enchoria* 12 (1984) 3, for example, I suggested since the answer to the question ἐν ποίᾳ ψήφῳ; is always τὸ X τῶν Y, then a ψῆφος must be τὸ X τῶν Y. Whereas P. Sijpesteijn, *op. cit.*, defended the genitive ψήφων (the tablet he edited contained that word with no preposition preceding it), I suggested in *BSAC* 26 (1984) 19ff., that one should emend ἐν ψήφων to ἐν ψήφοις, cf. Aeschylus, Agam. 570: τί τοὺς ἀναλωθέντας ἐν ψήφῳ λέγειν. Yet in view of Euripides, Rhesus 309-10: στρατοῦ δὲ πλῆθος οὐδ' ἂν ἐν ψήφου λόγῳ θέσθαι δυναίμην, an ellipse — ἐν (λόγῳ) ψήφων — also becomes conceivable.

Pap. ερχεται: The same phrasing on page 1 lines 24-25.

τὸ γ′′ $\overline{\text{ιδμβ}}$ (τάλαντα) Σ. πόσου τὸ $\overline{\text{μβ}}$ μέρος;
οὕτω ποίει· ἐν ποίῳ ψήφῳ ἔρχεται γ′′ $\overline{\text{ιδμβ}}$;
ἀπὸ ζ τῶ(ν) γ τὸ $\overline{\text{ζ}}$ γ′′ $\overline{\text{ιδμβ}}$. κ<ρ>άτησο(ν) τὸ ἔσ-
χατον μόριο(ν) $\overline{\text{μβ}}$ τὸ γ′′ $\overline{\text{ιδμβ}}$ γί(ν.) $\overline{\text{ιη}}$
τὰ προκ(είμενα) τάλαντα παρ(ὰ) $\overline{\text{ιη}}$ γί(ν.) (τάλαντα) ια<ϑ′> < χξϛβ'
> τὸ $\overline{\text{ιϑληνζος}}$ (δηναρίων) (μυριάδες) φ. πόσου μὲν τὸ $\overline{\text{ιϑ}}$
λη καὶ $\overline{\text{νζ ος}}$; οὕτω ποίει· κράτησον
τὸ ἔσχατον μόριον, τοῦτ' ἔστιν $\overline{\text{ος}}$,
τὸ $\overline{\text{ιϑ}}$ληνζος, γί(ν.) $\overline{\text{ηγ}}$′′. τρίπλωσον τὰς $\overline{\text{ηγ}}$′′
γί(ν.) $\overline{\text{κε}}$. διὰ τί "τρίπλωσον";

21 ἀπὸ ζ: "From (the table of) 7ths" (?), apparently in answer to the question of the preceding line. On page 1 line 26, where one would expect similar phrasing, it is lacking.
23 γί(ν.) has been written over.
Pap. <: Although it does not appear in the list of symbols and abbreviations for ἀριϑμός in T. Heath, *Diophantus of Alexandria* 1910² (repr. New York 1964), 32-37, that is obviously what it must mean here. Perhaps it is just a stylized *Hakenalpha* = ἀ(ριϑμός)
Pap.: χξϛβ': l. υξϛβ'
24ff. Calculating with parts: if 19' 38' 57' 76' of an unknown number (n) of myriads of denarii is 500, then how many are the 19' 38'? How many are the 57' 76'? D. Fowler points out that 19' 38' 57' 76' is the 19' of 1 2' 3' 4', thus the 19' of the 12' of 12+6+4+3, thus the 228' of 25. This would correspond to a result (ἐν ποίῳ ψήφῳ ἔρχεται) of 20. The scribe does not attempt to calculate this information or to use it, perhaps because the expression 19' 38' 57' 76' is not to be found in any standard division table. It should be emphasized that 228 occurs nowhere in this problem, not even implicitly. 25 only occurs incidentally. The scribe does not reduce the expression to anything resembling a common fraction.
19' 38' 57' 76' of 76 gives (4+2+1+3'+1 =) 8 3'.
19' 38' 57' 76' of n = 500
19' 38' 57' 76' of 76 = 8 3'

19-23 "3' 14' 42' (is) 200 talents. How much is the 42' part? Do it thus. In what ψῆφος does 3' 14' 42' go? (It is) from (the table of) 7ths. 3/7 is 3' 14' 42'. Take the last part, 42'. 3' 14' 42' (of 42 is) 18. (Dividing) the aforementioned (200) talents by 18 equals 11 9' talents. Number (of talents) 466 2/3".

24ff. "19' 38' 57' 76' (are) 500 myriads of denarii. How much are 19' 38' and (how much) are 57' 76'? Do it thus. Take the last fraction, that is 76. 19' 38' 57' 76' (of 76) is 8 3'. Triple 8 3', equals 25. Why "triple"? Because we have (the) 3'. Likewise, triple again the 500 myriads of denarii, (equals) 1500 (not 6500!). Divide this by 25, equals 60. Quadruple 60, equals 240. Half of this is 120. Again the 3' of 240 equals likewise 80, (that is) 80 myriads of denarii. Again, one quarter of (240 is) again 60 myriads of denarii, so that (19' is) 240 (myriads of denarii), 38' is 120 myr. den., 57' is ... (80 myr. den.) and 76' is 60 myr. den. Add them all together: 500 myr. denarii".

Hence, 76' of n is 8 3' of 500.
Triple, because of the 3', to get the 25' of 1500, or 60.
Thus the 19' of n is $4 \times 60 = 240$.
38' of n is 1/2 of 240 = 120
57' of n is 1/3 of 240 = 80
76' of n is 1/4 of 240 = 60
The check is to add them: 19' 38' 57' 76' of n = 500, or
19' 38', that is 240 + 120 = 360
57' 76', that is 80 + 60 = 140
500

24 λη: The eta has been re-written. The etas in this fraction are consistently different from other etas in these mathematical texts and comparable to the etas on p. 1 line 18.
25 κράτησον: "take"; cf. Heron, *Geometrica* and *Stereometrica*: κράτει.
27ff. τρίπλωσον τὰς $\overline{\eta}$γ// ... διότι ἔχομεν τὸ γ//: Cf. K. Vogel, *Ein byzantinisches Rechenbuch d. frühen 14. Jh.* (Wiener Byzantinische Studien 6), Vienna 1968, p. 106 no. 88.5: ποίησον μ ἐπὶ ια, γίνονται υμ. διατί δὲ ἐπὶ ια; διὰ τὸ ἀριθμὸν εἶναι τὸ ιβ// κβ// λγ// μδ// τῶν ια.
28 διὰ τί "τρίπλωσον"; Otherwise, one has to read διὰ τί τριπλοῦμεν/τριπλοῦται;

Extant page four

→

διότι ἔχομεν̣ [τὸ] γ´´. πάλιν ὁμοίως τρίπλω[σον]
τὰς (δην.) (μυρ.) φ <γί(ν.)> ϛ̣φ̅. ταῦτα μέρισον παρὰ τὸν
κ̅ε̅ γί(ν.) ξ̄. τετράπλωσον τὰς ξ̄ γί(ν.) Σ̅μ̅
τοῦτον τὸ 𐅵´ γί(ν.) ρκ. πάλιν ὁμοί(ως) τὸ γ̄´´ τῶ(ν) Σ̅μ̅
γί(ν.) ὁμοί(ως) π̄, (δην.) (μυρ.) π̄· πάλιν ὁμοίως τὸ τέταρτον
τ̣ῶν̣ [Σ̅μ̅ γί(ν.)] πάλιν ὁμοί(ως) (δην.) (μυρ.) ξ̄. ὥστε
[τὸ ιθ (δην.) (μυρ.)] Σ̣μ καὶ τὸ λη (δην.) (μυρ.) ρ̅κ̅. ὥσ<τε> τὸ <ν>ζ̄
[. . . (δην.) (μυρ.) π κ]αὶ τὸ ο̅ϛ̅ (δην.) (μυρ.) ξ· σύνθες
ὁ̣μ̣οῦ̣ ἐπὶ τὸ αὐτὸ γί(ν.) (δηναρίων) (μυριάδες) φ̄.

2 l. ᾿Αφ: The same mistake — ϛ instead of ᾿Α — appears on p. 3 line 4.
l. παρὰ τῶν.

3 Pap. τετρα πλωσον in order to avoid the kollesis.
τετράπλωσον: *addendum lexicis.* In connection with the appearance of this new word, formed on the analogies of διπλόω, τριπλόω, ἑξαπλόω it should be noted that the verbs διπλήζω, τριπλήζω, πενταπλήζω, ἑξαπλήζω, ἑπταπλήζω and δωδεκαπλήζω of *P. Achmim* (*s.v.* p. 13) almost a century after their appearance still have not found their way into any of our lexica.

4 l. τούτων.

5 After the first pi is something resembling chi. In fact, it is the denarius-symbol two of whose cross-strokes have been written practically over each other.

8-9 σύνθες ὁ̣μ̣οῦ̣: This reading seems preferable to ὁμοί(ως). The expression occurs in Heron, *Geometrica.*

JOHN 10:7-13:38 IN SUBACHMIMIC

WOLF-PETER FUNK
RICHARD SMITH

INTRODUCTION

1. The manuscript

The excerpt from the Gospel of John (chapters 10 to 13) preserved in the manuscript Chester Beatty ac. 1390 is of some rare interest to Coptic philology for at least two reasons. One is its particular "version" and linguistic form (supplying some valuable information in terms of dialectology), and the other is its particular medium — or the kind of manuscript that preserves this version. The "version" itself is not new: it is virtually the same as that known from the famous codex of the Gospel of John in Subachmimic Coptic published by Sir Herbert Thompson (1924), which largely translates the Sahidic version, notwithstanding a number of different readings. Even the actual type, or dialect, of Coptic in which this version is presented here, is not very different from that of the other codex. The nature of the two manuscripts, however, could hardly be imagined to be more different. While Thompson's manuscript is a first-class literary manuscript — a genuine biblical codex copied in calligraphy (a fact which also has favourable implications as to its orthography) — the manuscript published here is an excerpt written in a casual hand, apparently undecided between a semi-uncial and an almost fully cursive style, and preserved on papyrus leaves which, even if they formed parts of a codex at the time of inscription, never belonged to a biblical codex in the usual sense.

For some unknown reason, an unknown person (probably of the fourth century) used one or several papyrus quires partly inscribed with mathematical exercises to copy, after drawing a line across the width of the page below the last mathematical problem, the text of the Gospel of John beginning just before the end of 10:7. He continued copying, as it seems, at least up to some place in chapter 13, but probably further than that, possibly the whole rest of the gospel. The fact that he started his copying effort just in the middle of a sentence and continued through several chapters suggests that he was not copying a distinct part (or several such parts) of the gospel in the sense of producing "excerpts" determined by deliberate choice. Rather, he may have continued to copy the text just where he had stopped at a former time (whether in the same or another manuscript) — in this case he might have been interested in the actual text and aimed at a complete copy of the gospel. But he might as well have used the text merely for scribal practising, in which case his beginning may have been determined by the beginning of the page he was copying from. Whatever may have been the reasons for his work — it appears to be either a pre-editing or a post-editing effort, or both, but not in itself (as in the case of Thompson's codex) a final editing. One may say this is a literary text preserved by a non-literary document.

The scribal hand, in spite of the relatively large number of errors and irregularities, seems to be a professional hand. It strongly resembles a type of relatively upright cursive handwriting that is sometimes found in non-literary documents of the ancient period. To give an example, two Greek papyri containing letters which have been found among the cartonnage in the cover of Nag Hammadi Codex VII (no. 78, i.e. VII 20[c]/21[c], and no. 79, i.e. VII 23[c]/24[c], both before the middle of the fourth century) witness a hand not unsimilar in style to the present manuscript. This style, and the fluency of its performance, suggest a scribe experienced in writing documents, though not literary codices.

There is a considerable degree of inconsistency in the scribe's policy of arranging the column and lines of a single page. The borderline of the left-hand margin, where extant, can often be seen to be uneven, sometimes extremely oblique (most notably on p. 7*, probably also p. 6* if the present reconstruction of line beginnings is correct). In general, the available space is more crowded with lines, and the lines are more crowded with letters, and longer, in the upper part of a page than they are in the lower part. (The decrease in width of the column is, again, most obvious in the case of p. 7*, but it also applies to most of the other pages). Though sometimes found even in relatively good calligraphical manuscripts, such phenomena as just mentioned certainly do not enhance the aesthetic quality of hand-written pages.

What is more interesting, however, is the impression one may gain from comparing different pages with one another and following up the scribe's writing behaviour from page to page. Facing pages are of course always more likely to display a similar layout than are consecutive recto/verso pages. But there can be no doubt that on the first two pages (pp. 4*-5*) the writing is on the whole much more cursive than in the rest of the manuscript. The switch in style from page 5* to page 6* is so striking that at first sight one may doubt their being written by the same hand. However, the identity of the hand is evident enough from the recurrent presence of certain fully cursive patterns (involving particular types of ligature) on the less cursive pages, and vice versa — elements of the same type of "semi-uncial" writing as found on the later pages do already appear in places on the first two pages. In fact, there is never any absolute difference in terms of actually occurring shapes of letters but rather a quantitative shift in the predominance of the different styles or patterns of writing used on the various pages.

More precisely, though perhaps in a somewhat oversimplifying manner, one may say that the scribe initially, in the first three lines of his part of p. 4*, employs what may be termed 'his best variety of a semi-uncial style', but in the next two or three lines gets more and more lost in cursive style; from the seventh line onward, he has definitely fallen into cursive writing, from which he does not emerge until the beginning of page 6*. Then he succeeds in performing his best variety of a semi-uncial throughout page 6* and the following pages. (Could it be due to this particular endeavour that he loses control of the overall shape of the column on these pages?) From time to time, however, he suffers from severe relapse into cursive style, for instance, in the central part of p. 9*, in the upper central part of p. 10* and some central lines of p. 11*. He resumes and maintains, perhaps even slightly improves, his best variety throughout pages 12* to 14* (or 15*), while the fragments of the last preserved leaf (especially on p. 16*) seem to signal another fall-back into cursive writing (combining with uneven lines) in the lower parts of the pages. If the impression of such a gross development is justified, this may suggest a strong endeavour on the part of the scribe to improve his style of writing. It is impossible to say, however, whether this was his primary (or sole) intention in copying the text.

As far as dating is concerned, there is no reason to assume that the gospel text was copied much later than the mathematical exercises preceding it in the manuscript. As these can be dated in the first half of the fourth century with some certainty, the assumption of contemporaneity of the two parts of the manuscript may be safer than any palaeographical guess regarding the Coptic part alone — which, however, would probably point to the first half of the fourth century as well.

2. Spelling and Orthography

To tell right from wrong is especially hard in the case of the present manuscript. There can be no doubt that the scribe made numerous mistakes and that he noticed and corrected only some few of these while the rest escaped his attention. His usual spelling practice, however, reveals such a lot of bizarre features that particular caution may be warranted in judging what should be regarded as actual errors or slips of the scribe. Although it seems almost impossible to obtain a clear picture of the kind of orthography he was aiming at, some clues may be provided by taking a closer look at the slips of the pen that the scribe recognized as such and corrected himself. Here is a complete list of these corrections (quoted with page and line):

(a) Substitution of a letter by superimposing it on the wrong one
(b) Substitution of a letter by writing it above the line and deleting the wrong letter(s)
(c) Insertion of one or several letters by writing it/them above the line near the place of omission

		before correction	after correction
(a)	4:16	(identity of letters unclear; see *infra*, apparatus)	
	5:23	ⲁⲛⲁⲕ	ⲁⲛⲁⲛ
	5:25	ⲛⲉϥ	ⲛⲉⲩ
	6:26	ⲡⲁϫⲉⲩ (ⲩ incomplete)	ⲡⲁϫⲉϥ
	8:23	ⲉⲩⲛⲱ(ⲛⲉ)	ⲉⲩⲛⲟⲩⲱⲛⲉ
	10:5	ⲉϥⲛⲁϭⲁⲡϥ	ⲉⲩⲛⲁϭⲁⲡϥ
	14:8	ⲁϥⲛⲁϫⲙⲁϥ	ⲁϥⲛⲁϫⲙⲁⲩ
(b)	9:20	ⲛⲧⲁⲣⲉϥϫⲉⲛⲉⲉⲓ	ⲛⲧⲁⲣⲉϥϫⲉⲡⲉⲉⲓ
	11:18	ϫⲉⲉⲩ	ⲛⲉⲩ
	16:22	ϯⲛⲁϣⲟⲩⲁⲕ	ϯⲛⲁϣⲟⲩⲁϩⲧ
(c)	4:11	ⲡⲉⲛⲉⲥⲁⲩ	ⲡⲉⲛⲙⲁⲛⲉⲥⲁⲩ
	4:28	ϯⲛⲁⲥⲁϩⲟⲩ	ϯⲛⲁⲥⲁⲩϩⲟⲩ
	5:11	ⲁⲩⲱ(ⲧⲉ)	ⲁⲩⲕⲱⲧⲉ
	5:21	ⲙⲧⲉⲗⲁⲩⲉ	ⲙⲧⲉⲧⲙⲗⲁⲩⲉ
	6:6	ϫⲡⲁⲉⲓⲱⲧ	ϫⲉⲡⲁⲉⲓⲱⲧ
	10:30	ϩⲛϩⲉⲛⲃⲛⲉ	ϩⲛϩⲉⲛⲃⲛⲛⲉ
	12:10	ⲁⲛⲁⲕⲟⲩⲧ	ⲁⲛⲁⲕϩⲟⲩⲧ
	14:21	ⲁⲗⲗⲁⲧⲟⲩⲃⲁⲉⲓⲧ	ⲁⲗⲗⲁϥⲧⲟⲩⲃⲁⲉⲓⲧ

Furthermore, a passage comprising several words has been deleted in 5:26/27 by encircling the respective parts of these lines (see *infra*, apparatus).

A first observation may be made with regard to the effectiveness of the scribe's correction practice. It is remarkable that in at least four cases the result of his correction falls short of a satisfactory reading: in 4:11 he changes the correct reading "the sheep" into an awkward "the shepherds" (note that he does not change the Neg. Perf. base ⲡⲉ- to read either ⲙⲡⲉ- or ⲛⲡⲉ-); in 5:21 he correctly inserts ⲧⲙ but does not change ⲙⲧⲉ- into ⲛⲧⲉ-; in 5:23 his correction is entirely based on error, involving the omission of ⲁⲛⲁⲕ ⲙⲛⲡⲁⲉⲓⲱⲧ by homoioteleuton;

and in 5:26/27 his correction goes too far, deleting even words which are not due to dittography (see *infra*, apparatus). Although incomplete corrections sometimes do occur even in good MSS, these examples may suggest a rather low degree of attentiveness and/or skill on the scribe's part.

The two types of error most frequently noticed and corrected by the scribe clearly involve wrong pronominal suffixes (especially -ⲩ and -ϥ; cf. also p. 14:8 where final -ⲩ is part of ⲙⲁⲩ 'water') and the omission of one or two letters. Wrong pronominal suffixes are also found without correction in 4:16; 6:24, 33, 37; 8:34 (probably not wrong suffix, but incomplete letter); 9:30, 31 bis; 11:17; 12:17; 13:22. Omission of the suffix occurs in 8:11 and 16:18. (Suffixes different from the ones found in T [=Thompson] may, however, be correct in 8:32; 12:13; 13:13.) Concerning personal pronouns in general, it may also be noted that the scribe twice writes ⲧ instead of correct ϯ, viz., p. 7:32 for the First Present preformative of 1st sing., and p. 13:22 in what should read ⲡⲉϯⲛⲁ-, although the scribe's ⲡⲉⲧⲛⲛⲁ- might have a possible (mis)interpretation in terms of the 1st pl.

There are numerous cases of omission of one or two letters which were not corrected by the scribe. Apart from haplography, these include 4:17 ⲟⲩϩ<ⲟⲩ>ⲟ, 5:15 <ⲉ>ϯⲉⲓⲣⲉ, 6:7 ⲛⲉⲩϣⲓ<ⲛⲉ>, 6:30 ⲙⲛ<ⲧ>ⲥⲛⲁⲟⲩⲥ, 8:3 <ⲁ>ⲥⲧⲱⲱⲛ, 8:11 ⲁⲥⲛⲁϫ<ⲥ>, 8:17 ⲛ<ϭⲓ>ⲓ̅ⲥ̅, 8:19 ⲙⲁ<ⲉⲓ>ⲉ, 8:31 ⲟⲩⲁ<ⲉⲓ>ϣ, 9:25 ⲥⲁ<ⲩ>ϩⲟⲩ, 9:30 ⲡⲟⲗⲓ<ⲥ>, 10:24 <ⲁⲩ>ϫⲓ-, 11:12 <ⲁ>ⲟⲩⲱϣⲧ, 11:26 ⲟⲩ<ⲛⲟⲩ>ⲉⲉ. Other cases of plain misspelling, sometimes involving metathesis, are 6:5 ϩⲁⲛⲃⲏⲩⲉ for ⲁⲛϩⲃⲏⲩⲉ, 6:16 ⲁⲥϥⲛⲧ- for ⲁⲥϥⲁⲧ-, 7:3 ⲡⲉⲉⲓⲛ̄ⲧⲁⲕⲉⲛ for ⲡⲉⲉⲓⲛ̄ⲕⲁⲧⲕⲉ, 10:33 ⲟⲩⲉⲉⲓϣ for ⲟⲩⲉⲓⲱ, 8:20 ⲛⲁⲃⲉⲗ for ⲁⲛⲃⲉⲗ. Omission of the letter ϩ (as corrected in 12:10) occurs at least in 6:3 ⲛ̄<ϩ>ⲃⲏⲩⲉ (cf. also the misspelling, just quoted, in 6:5), while 5:22 ⲉϥⲓϫⲛ- and 4:12 ⲉⲓⲧⲟⲟⲧϥ may (?) have a chance of being intended by the scribe as true variant spellings of ⲉϥϩⲓϫⲛ- and ϩⲓⲧⲟⲟⲧϥ, respectively. Omission of an entire word through haplography occurs twice across the borderline between sentences: 5:18 and 11:16. For other cases of apparent omission, in particular that of initial nasals, see the discussion below.

Somewhat less frequent are redundant letters that are clearly due to scribal error; cf., e.g., 8:20 ⲙⲡ{ⲣ̄}ⲃⲁⲗⲉ, 9:20 ⲛⲧⲁ{ⲡⲉ}ϥϫⲉ-, 10:11 ⲥⲟⲩⲛ{ⲥ}ⲧⲥ, 10:16 ϣⲙⲧϣ{ⲱ}ⲉ. Dittography of a whole group of letters occurs in 6:36/37 ⲉⲓⲛⲁⲧⲟⲩⲛ{ⲟⲩⲛ}ⲁⲥϥ and 12:9 ϯⲛⲟⲩ{ⲛⲟⲩ}; for the dittography of (one-letter or two-letter) glide graphemes see the discussion below. The letter ϩ is sometimes redundant: neighbouring a glide in 6:13 ⲛⲉⲩ{ϩ}ⲛ-, but mostly replacing a phonetic alif (glottal confusion — the complementary phenomenon to h-dropping), as in 7:25 ⲙ̄ⲫⲁⲉ {ϩ}ⲛϩⲟⲟⲩ, 10:16 {ϩ}ⲛϩⲏⲕⲉ, (for ⲛⲛϩⲏⲕⲉ), 11:25 and 12:21 ⲉϥ{ϩ}ⲛⲛⲁ (the stative ⲛ̄ⲛⲁ "be going to go"). For dittography of consonants and vowels see further below.

All this contributes to the general impression that the scribe may have been rather incompetent or, at least sometimes, extremely inattentive — in either case, this record tends to diminish the weight of whatever features might be considered as true spelling variants in terms of the scribe's intention, and it certainly does not compel us to seek a rational explanation, in terms of phonetic, phonological, or morphological motivation, in every single case. However, the scribe's spelling and correction practice also reveals some outstanding features which deserve a closer look.

As a first thing, it is important to see that the scribe never makes any effort to change a ⲭ into the well-known Coptic letter ϫ. That is to say, what is perhaps the most salient spelling feature of this MS — the constant use of the Greek letter ⲭ in all cases where ϫ is expected — is a regular feature of the scribe's orthography. We thus have ⲭⲉ, ⲭⲉⲕⲁⲥⲉ, ⲡⲁⲭⲉ-, the verbs ⲛⲟⲩⲭⲉ, ⲥⲉⲭⲉ, ⲥⲁⲭⲛⲉ, ⲧⲟⲩⲭⲟ, ⲟⲩⲭⲉⲉⲓ, ⲭⲱⲕ, ⲭⲱⲕⲙ, ⲭⲓⲥⲉ, ⲭⲱⲣⲉ, ⲭⲁⲩ, ⲭⲓⲟⲩⲉ, the nouns ⲭⲁⲓⲉ, ⲭⲣⲁⲡ, ⲭⲁⲉⲓⲥ, the prepositions ⲁⲭⲛ-, ϩⲓⲭⲛ-. There can be no doubt about the identity of the letter ⲭ in these forms: the letter does not just lack the basic stroke of ϫ but is shaped like a typical Greek cursive *χ* and thus has not much in common with any form of Coptic ϫ. One might expect that in a MS that regularly uses the letter ⲭ to designate the phoneme /č/ — and thus to function, strictly speaking, as a grapheme <č> — there ought to be no room for the peculiar use of ⲭ, which is found elsewhere in early Coptic MSS outside Bohairic, as a monographic variant of the sequence <kh> (or ⲕϩ). This expectation, however, is not confirmed. If the scribe successfully avoided the letter ⲭ in spelling ⲙⲁⲕϩ in 8:22, he had already spelled the same form ⲙⲁⲭ in 8:15 (in both cases T has ⲙⲁⲭ). Thus the letter ⲭ is potentially ambivalent in the present MS: it may stand for the sequence ⲕϩ, just as ⲫ stands for ⲡϩ and ⲑ for ⲧϩ (see Index), although it is primarily used to represent the phoneme /č/.

This situation is clearly different from any other Coptic MS that makes use of ⲭ for ϫ. The nearest parallel is perhaps the series of Coptic glosses found in the margin of Chester Beatty Papyrus VII

(Isaiah; a Greek MS). There the use of ⲭ instead of ϫ is as exhaustive as it is in the present MS, but so is also the use of ⲥ for ϣ, of ⲃ for ϥ, of ⲧⲓ for ϯ, and the non-use of ϩ. In other words, the writer of these glosses does not use any non-Greek letters. In the present MS, on the other hand, all non-Greek letters of the standard Sahidic (and Subachmimic) alphabet are known and regularly used, except ϫ. Furthermore, the Isaiah glosses are not a "text" strictly so called, and the idiom they display, though having much in common with Fayyumic dialects, cannot be said to represent any cultivated literary dialect. It seems quite plausible that these glosses are "the work of a Coptic-speaking owner of the papyrus, who wrote them ... for his private edification" (Crum 1937: ix), and that he did so "at a time when the five supplementary native letters were — at any rate, in his neighbourhood — still unfamiliar" (*ibd.* x-xi) — which it is not too hard to imagine, given the dating of that MS to the first half of the third century. (Note that all such considerations are not directly applicable to the present MS.) All other MSS that have so far been known to use ⲭ for ϫ (among them, most notably, the quasi-Achmimic Psalm fragment ed. Crum 1934, for which see Kasser 1981: 113f.; see also Crum, *Dict.* 516a) do so only sporadically, and in most cases, probably by mistake.

The use of the letter ⲭ to represent /č/, in the present MS, probably has nothing to do with the substitution of ϫ for ⲭ in certain Greek words, especially when preceding ⲓ, in several early MSS (among them, T). The latter phenomenon clearly has a phonological foundation (cf. Thompson 1924: XIX) and presupposes the distinction between ⲭ and ϫ.

It seems possible that the scribe's non-acquaintance with the Coptic letter ϫ may be reflective of a historical situation where this particular letter and thus the distinction between ⲭ and ϫ, in the scribe's milieu, was still unknown (in this case the phenomenon might be comparable to the merger of the letters ⲯ and ϯ in some early MSS). This is the interpretation favoured by R. Kasser: "un autre exemple de maladresse archaïque dans l'écriture copte (chez certains scribes du moins)" (personal letter, July 23, 1985). For a more elaborated account of his opinion see Kasser 1986b: 227, where he proposes to relate the Coptic letters ⲭ and ϫ to slightly different graphic variants for *ḏ* in Demotic and hypothetically assumes their indifferent usage "au premier stade vieux-copte". While this seems possible, it should be borne in mind that the present document itself is not a specimen of "vieux-copte" but clearly "copte proprement dit", i.e., it presupposes the existence and usage of one of the full-fledged Coptic literary dialects of the fourth century. Notwithstanding any potential historical trajectories, it also seems possible that the use of ⲭ instead of ϫ is just due to the individual lack of expertise and familiarity with normal literary Coptic spelling (already current at this time) on the part of the scribe. There are some other indications that he may not yet have learned all of his Coptic spelling lessons, or not all of them equally well, when he wrote this text.

Perhaps the most striking feature of the scribe's correction policy is the fact that he never emends any of the following three categories of incorrect or, at least, unusual spellings, which are so characteristic of his performance: (1) omission of the stressed vowel in certain words, (2) omission of initial (syllabic) ⲛ- or ⲙ-, (3) dittography of vowels, glides, or consonants at word or morph boundaries. These categories, therefore, even though the second and third of them may represent typical features of non-literary scribal practice, deserve some special attention.

The first of these categories, viz., omission of the stressed vowel, is found in the following forms (for references see the Indices): ⲥϩ for ⲥⲁϩ 'teacher', ⲛⲁϣϥ for ⲛⲁϣⲱϥ (from ⲛⲁϣⲱ⸗ "be numerous"), ϣⲥ for ϣⲱⲥ "shepherd", and a number of cases where ⲱ is omitted after ⲟⲩ: ⲟⲩⲧ, ⲟⲩϣ, ⲟⲩϣⲃ and ⲟⲩϩ. Of these, the last-mentioned spellings may relatively easily be dismissed as mistakes (although it should be noted that the presence of ⲟⲩ here always provides for a possible substitute of the stressed vowel and thus allows a phonetic interpretation in terms of assimilation): ⲟⲩⲧ occurs only once (but three times ⲟⲩⲱⲧ), ⲟⲩϣ once for the st. abs. of the verb ⲟⲩⲱϣ (besides ⲟⲩⲱϣ- for the st. cstr.) as well as once for ⲟⲩⲁⲉⲓϣ "time", ⲟⲩϣⲃ only twice (as against six times correctly ⲟⲩⲱϣⲃ, and ⲟⲩϩ only once. The mere occurrence of the letters ⲱ and ϣ in the neighbourhood appears to have prompted also other mistakes: one of these letters is redundant, respectively, in ϣ{ⲱ}ⲉ 10:16 and ⲟⲩⲱ{ϣ}ⲛϣ 4:21 (the latter, however, may be a mistake for ⲟⲩⲱⲱⲛϣ; wrong ϣ for ⲱ also in 10:33). With some scribes, omission of the stressed vowel typically occurs in the neighbourhood of glides; cf., e.g., ⲟⲩ<ⲟ>ⲉⲓⲛ, ⲧⲁⲟⲩ<ⲟ>ⲉⲓ, and ⲉⲣ<ⲟ>ⲉⲓ in the Berlin MS of Psalms (ed. Rahlfs 1901: 39). Even the form ⲥϩ, attested only once, might be a mistake. In the remaining two cases, however, it seems difficult to dismiss these awkward spellings as plain mistakes: ⲛⲁϣϥ is spelled this way in all three instances of the st. pron. being used (10:21, 28; 11:23; but cf. the st. nom. ⲛⲁϣⲉ-), and ϣ̅ⲥ̅ occurs three times (once with and twice without the supralinear stroke) in the expression ⲡϣ̄ⲥ̄ ⲉⲧⲛⲁⲛⲟⲩϥ "the good shepherd", whereas the normal spelling ϣⲱⲥ is used outside this particular syntagm. The use of a quasi-construct form in this syntagm may call to mind the somewhat similar use of ⲣⲙ̄ in the pattern ⲡⲣⲙ̄ ⲉⲧ- in P. Bodmer VI (ed. Kasser 1960). It should be noted, however, that ⲡⲣⲙ̄ ⲉⲧ- in that MS is always followed by a Bipartite verbal predicate and is mostly found to translate Greek adjectives, as the definite counterpart of ⲟⲩⲣⲙ̄ ⲉϥ- (for which see also Jn 9:8, 25 in T), while normal ⲡⲣⲱⲙⲉ ⲉⲧⲛⲁⲛⲟⲩϥ is used in Prov 14:33 to translate ἀγαθὸς ἀνήρ. It may also be noted that in P. Bodmer VI the adverb ⲁⳉⲟⲩⲛ has a variant spelling ⲁⳉⲛ̄ when it immediately precedes the preposition ⲁ-, ⲁⲣⲟ⸗ (Prov 5:10; 7:27; 9:18b; without the stroke 11:2), while it is always spelled ⲁⳉⲟⲩⲛ when it is the final constituent of a verbal syntagm.

Notwithstanding the somewhat peculiar case of ϣⲥ, however, both ϣⲥ and ⲛⲁϣϥ, perhaps also ⲥϩ and ⲟⲩϩ, may deserve being taken seriously — either as reminiscent of a consonant-based writing system (a "demoticizing" feature of orthography) or as reflective of actual speech habits. If at least some of these can pass as true spelling variants, they may find their counterpart in the spelling ⲉϣⲱⲡⲉ in those cases where it stands for ⲉϣⲡⲉ- (see below, at the beginning of the next chapter).

As to the second of the above-mentioned categories, omission of an initial syllabic nasal occurs both with morphemic and non-morphemic nasals, and it occurs both in the immediate neighbourhood of another nasal (so that the "mistake" may be categorized as haplography) and with no other neighbouring nasal.

Omission of non-morphemic initial syllabic nasal
(For references see Indices).

(a) Preceding another nasal
ⲙⲉⲩ, ⲙⲁ⸗ (for ⲙ̄ⲙⲉⲩ, ⲙ̄ⲙⲁ⸗)
ⲛⲁ, ⲛⲏⲩ (for ⲛ̄ⲛⲁ (?), ⲛ̄ⲛⲏⲩ (?))
NB: ⲛⲁ and ⲛⲏⲩ, being the standard spellings of these verbs in Sahidic orthography, may of course be considered "normal". If looked upon from the standpoint of T, however, the current forms ⲛ̄ⲛⲁ and ⲛ̄ⲛⲏⲩ might be regarded as "normal" spellings for the present MS, too, even if the variant ⲛ̄ⲛⲁ is only attested as {ϩ}ⲛⲛⲁ. (Note that in T there is indeed a scribal correction \ⲛ̄/ⲛⲏⲩ Jn 4:25, which would seem inconceivable for the scribe of the present MS.)

(b) Preceding a non-nasal consonant
conjugations ⲡⲉ- (Neg. Perf.), ⲡⲁⲧⲉ- (Neg. Completive)
conjugations ⲧⲁⲣⲉ- (Temp.), ⲧⲁ- (Second Perf.)
[Conjunctive ⲧⲉ-, ⲕ, ⲧⲉⲧⲛ]
personal pronouns ⲧⲁⲕ, ⲧⲁϥ
prepositions ⲧⲉ-, ⲧⲟⲟⲧ⸗, ⲥⲱ⸗, ϩⲏⲧ⸗, ϭⲓ-
adverb ϩⲣⲏⲓ
NB: Initial ⲛ̄- in "compound" prepositions and adverbs may of course be analyzed as "morphemic". — As to the Conjunctive, there is some variation between forms with and without ⲛ even in T, with the latter variants probably being an Achmimicism.

For some of these phenomena, in particular ⲙⲉⲩ and ⲙⲁ⸗, the closest parallel apart from non-literary texts seems to be the London Didache (cf. Schmidt 1925:83), another example of a literary text in a non-literary MS (there also ⲡⲉⲣ- for the Neg. Imp. ⲙⲡⲉⲣ-).

Omission of morphemic initial syllabic nasal

(a) Following another nasal (final to the preceding word) and thus possibly, though not necessarily, omitted through "technical" haplography
5:16 ϩⲛⲡⲣⲉⲛ ⲡⲁⲉⲓⲱⲧ
6:3 ⲉⲛ ⲡⲁⲉⲓⲱⲧ
10:32 ϩⲛⲡⲣⲉⲛ ⲡⲭⲁⲉⲓⲥ
(b) Preceding another nasal (onset of the following syllable) and thus possibly omitted through "phonetic" haplography (i.e., the monographemic representation of consecutive identical phonemes in different syllabic functions)
6:14 ⲁⲡϯⲙⲉ ⲙⲁⲣⲓⲁⲙ
6:26 ⲡⲁⲭⲉϥ ⲛⲉϥⲙⲁⲑⲏⲧⲏⲥ
7:13 ⲁⲃⲁⲗ ⲛⲓⲟⲩⲇⲁⲉⲓ
9:23 ⲧⲡⲁⲥⲭⲁ ⲛⲓⲟⲩⲇⲁⲉⲓ
(For the latter two cases cf. Quecke 1984a: 54.)
Inside a phonological word (pl. def. art.):
6:27, 37 ⲛϭⲓ-ⲙⲁⲑⲏⲧⲏⲥ
(c) Preceding a glide: this case, which is so common in good early MSS, particularly in connection with prevocalic ⲟⲩ- (cf. Emmel 1981: 142 n. 5; Quecke 1972: 39, 45f.; Rahlfs 1901: 36) is here represented only once by its standard example ⲟⲩⲱⲧ (9:19 ⲟⲩⲣⲱⲙⲉ ⲟⲩⲱⲧ, but ⲛ̄ⲟⲩⲱⲧ in three other cases); the text

provides no occasion for words such as ⲟⲩⲁⲉⲓⲛ or ⲉⲓⲱⲧ in attributive construction.

(d) Preceding a non-sonorant consonant, and not following another nasal

4:25 ϯⲥⲁⲩⲛⲉ ⲡⲁⲉⲓⲱⲧ
4:27 ⲙ̄ⲙⲉⲩ ϩⲉⲛⲕⲉⲉⲥⲁⲩ
5:13 ⲉⲕϥⲓ ⲡⲉⲕϩⲏⲧ
6:8 ⲁⲡⲓⲕⲣⲟ ⲡⲓⲟⲣⲇⲁⲛⲏⲥ
8:2 ⲛⲉⲥ ⲭⲓⲟⲩⲉ
9:1 ϩⲛⲛⲟⲩⲛⲁϭ ⲥⲙⲏ
9:3 ⲙⲛⲛⲉϥⲟⲩⲣⲓⲧⲉ ϩⲉⲛⲕⲉⲣⲉⲁ
9:34 ⲁⲑⲓⲉⲣⲟⲩⲥⲟⲗⲏⲙ ⲧⲭⲱⲣⲁ
10:12 ⲙⲙⲁⲩ ⲡⲉⲥ<ϥ>ⲟⲩⲉ
11:19 and 12:17 ⲡϣⲏⲣⲉ ⲡⲣⲱⲙⲉ
12:12 ⲛⲉϥⲭⲱ ⲇⲉ ⲡⲉⲉⲓ
12:16 ⲛⲉϣ ϩⲉ

Inside a phonological word (pl. def. art.)

9:13 ⲛ̄ϭⲓ-ϩⲣⲱⲙⲁⲓⲟⲥ

Those cases where the lacking ⲛ- is the object marker (4:25, 27; 5:13; 12:12) may be counterpointed by the presence of a redundant object marker following the construct infinitive form in 10:26 ⲉⲩⲁⲙⲁⲟⲩⲧ-{ⲙ}-ⲡⲕⲉⲗⲁⲍⲁⲣⲟⲥ and, if -ⲟⲩ- is an assimilated ⲛ-, also in 12:11 ⲛⲧⲁ{ⲛ}ⲥⲁⲕ-{ⲟⲩ}-ⲟⲩⲁⲛ (but perhaps this is just a "technical" dittography, or an erroneous suffix). The spelling in 10:26 may be explained in terms of a "gliding" sonorant bridging the gap between adjacent occlusives (cf. also intrusive ⲛ in 8:8 ⲁⲩⲟⲩⲁϩⲟⲩ{ⲛ}ⲛ̄ⲥⲱⲥ, probably inserted to bridge the hiatus).

As the high frequency of "omissions" of initial nasals shows, these can hardly be "slips of the pen", and if we were to put them down to the carelessness of the scribe, this would be an extreme case. At any rate, the question remains: why did he care about initial nasals less than other things? In some cases, especially those involving non-morphemic ⲛ-, mere phonetic dropping of the initial nasal seems to be a possible explanation. But this explanation would cause severe problems when applied to the omission of morphemic ⲛ-. The *letter* ⲛ can be dropped — the *morpheme* cannot (except by actual mistake). At least in those cases where the morpheme signifies a "genitive" relation, but certainly not only there, it is indispensable. If the morpheme, in some such cases, was represented by graphic zero, what could this mean in terms of (morpho)phonology? Could it be that the scribe's spelling practice accounts for some sort of current phonetic realization to such a high degree that it interferes with any reasonable orthography? And what could this phonetic realization be like, hypothetically?

To make an attempt to answer these questions, it may be helpful to take a look at another case of n-dropping in the present MS, this time word-interior, in spelling the verb "live". The infinitive (or noun) is regulary spelled ⲱⲱϩ, once shortened to ⲱϩ at the end of a line (7:26), and the stative is spelled ⲁⲁϩ. According to widely accepted principles for the phonological interpretation of written Coptic, these forms represent /oʔh/ and /aʔh/, respectively. The presence of phonological aleph in these forms is further corroborated by T, which does not, however, drop the /n/: ⲱⲱⲛϩ and ⲁⲁⲛϩ, i.e., /oʔnh/ and /aʔnh/ (the same spellings are also used by the scribe of NHC I,*4* and XI,*1-2*). What is more, the forms used in the present MS are phonologically identical (except for /h/ ~ /x/) with those in P. Bodmer VI (ed. Kasser 1960), the only Coptic MS known so far whose writing system has a special grapheme <ʔ>, represented by the letter ⊥: its regular spelling of the verb "live" is ⲱ⊥ⳉ, which is paralleled by other cases of word-interior ⊥ such as ϣⲟ⊥ⲡ and ϫⲟ⊥⸗, and thus is most plausibly to be interpreted as /oʔx/. (In contrast to Kasser [1960: xx], I do not believe that the letter ⊥ is ever seriously used to signal /n/, or that ⊥ and ⲛ are constantly confused, though ⊥ and – may be; for a modified view on this issue see Kasser 1986a). The question may be left open whether this usage reflects a total change of [n] into [ʔ], or just an intrusive (or metathetic) /ʔ/ in a phonetic form [oʔnx] where the [n] was not felt to be salient enough to be represented graphically (and thus cannot count as phoneme).

As to the morphemic initial ⲛ̄- under discussion here, the usage of P. Bodmer VI may also provide an essential clue. While ⊥⸗ and ⲛ̄ⲧ(ⲉ)⸗ are found to be in perfect complementary distribution as allomorphs of the Conjunctive base (/ʔ⸗/ and /(ʔ)nt(e)⸗/, respectively), the allomorph ⊥- (i.e., /ʔ-/) is used, it seems, as a free variant of morphemic ⲛ̄- (note that ⊥- is not listed in Kasser's index apart from its occurrence in ⊥ⲧϩⲉ pp. 30:9 and 102:2; it is not frequent, but cf., for instance, pp. 9:3 [pl. def. art.] and 13:14 [object marker]). We may assume that the scribe of P. Bodmer VI based his spelling policy on a general phonological representation of this element as /ʔn-/, with the glottal stop being felt as a salient feature, whereas the nasal was subject to assimilation into any sort of sonorant, phonemically indistinct. In his spelling practice, then, he would tend to be inconsistent, having this element represented by ⲛ̄- (where /ʔ/ is implicitly represented) in most of the cases, but occasionally by ⊥-, choosing each time what he thought to be more salient. (The actual "full" spelling ⊥ⲛ̄- is extremely rare; it is found, for instance, in ⲙⲁ ⊥ⲛ̄ⲛⲓ p. 47:10, where the presence of vowels immediately preceding and following the element seems to have prompted a perfectly explicit phonetic spelling for what is morphologically ⲙⲁ ⲛ̄-ⲉⲓ, or /ma(ʔ) ʔn-i/, phonetically ['ma(ʔ).ʔn̩.'ni:]; the letter ⊥, in this case, cannot strictly be attributed to either ⲙⲁ or ⲛ̄-, since it is justified only in rendering a phonetic sequence).

It is imaginable that a scribe who experienced a similar linguistic situation to that of the scribe of P. Bodmer VI, but to whom no grapheme <ʔ> was available, may have been prompted, from time to time, to a zero representation (as usually in word-final position) of phonological /ʔ/, and that this automatically resulted in graphic zero for the morpheme {ʔn-} whenever the nasal was felt not to be salient enough to be represented graphically. It goes without saying that this consequence could never have been an acceptable solution in any system of literary Coptic orthography; rather, it would join the set of non-literary habits that take too much account of fluctuation in actual phonetics ("la parole") and thus always tend to spoil good orthography. There is some other evidence that the graphic representation of [ʔ], whether phonemically relevant or not, may have been a recurrent problem for the scribe of this MS.

The third category of "error" never corrected by the scribe involves the more general phenomenon of dittography and, to a smaller extent, haplography of single letters. As far as "consonant" letters are concerned, this applies almost exclusively to sonorants, and it is the ambivalent function of sonorants, glides, and glottal stops in the transfer of morphemes into syllables that seems to be the main cause of trouble here.

Consonant dittography/haplography

(a) Inside a lexical word or morph
Dittography, here, seems to be limited to the case of "belated", and thus wrong, decision with regard to the division of syllables between lines. Not uncommon even in good literary MSS, this phenomenon is here witnessed by spellings such as ⲧⲉⲥⲥⲱⲛ|ⲛⲉ 8:1/2 and ⲉⲧ|ⲧϥⲙⲁ<ⲉⲓ>ⲉ 8:18/19. The only case of word-interior consonant (sonorant) haplography has been corrected by the scribe: 10:30 ⲃ̀ⲛ́ⲛⲉ.

(b) At morph boundary (although inside a phonological word)

This case is comparable to the well-known double spelling of (syllabic) sonorants when they precede vocalic onsets (as in ϩⲛ̄ⲛⲟⲩ- for ϩⲛ̄-ⲟⲩ- or in ϩⲛ̄ⲛⲙ̄- for ϩⲛ̄-ⲙ̄-), which also occurs in the present MS (e.g., in ⲁⲭⲛⲛⲟⲩⲥⲏϭ 12:2, ϩⲛⲛⲉϣ 12:13). With the double spelling of sonorants always reflecting disyllabic function, it here takes account of the role of the sonorant in providing a consonantal onset to the second syllable. In addition to this, the present MS occasionally spells double nasals also in forms where the first element ends in a vowel and the second begins with a sonorant, with the doubling of the sonorant letter probably reflecting the additional coda role of the sound with regard to the preceding syllable, in normal allegro speech.

4:18 ϩⲁ{ⲛ}ⲛⲉϥⲉⲥⲁⲩ
5:4 ⲉⲧⲃⲉ{ⲛ}ⲛⲉⲉⲓⲥⲉϫⲉ
10:11 ⲉⲛⲛⲁϣⲉ-, 10:21 ⲉⲛⲛⲁϣϥ (both with circ. ⲉ-)
11:24 ⲡⲉⲧⲛⲁ{ⲙ}ⲙⲉⲥⲧⲱ-ⲧⲉϥⲯⲩⲭⲏ
14:12 -ⲉⲓⲁ{ⲛ}ⲛⲁⲟⲩⲣⲓⲧⲉ
14:21 -ⲓⲁ{ⲛ}ⲛⲉϥⲟⲩⲣⲓⲧⲉ (both ⲉⲓⲁ- 'wash')

A different case is exemplified by ⲡⲣ̄ⲣⲟ 4:12 (for ⲡⲣⲟ) and ⲉⲧⲛⲛⲁ- 13:18, 19 (for ⲉⲧⲛⲁ-). Here the scribe seems to extend the initial syllabic sonorant pattern, which is orthographically peculiar to similar lexemes (ⲣ̄ⲣⲟ and ⲛ̄ⲛⲁ), to the forms ⲣⲟ and ⲛⲁ-. To the scribe, this may have been reflective of an exaggerated lento pronounciation while maintaining close juncture.

Haplography, under this category, is only found in a single case, involving a non-sonorant consonant:

6:17 ⲡⲉⲥ<ⲥ>ⲁⲛ

(c) At the boundary between phonological words: doubling of word-final consonant

6:20 ⲉϥϣⲟⲡ{ⲡ}ⲉⲛ
7:27 ⲕⲁⲛ{ⲛ}ⲉϥϣⲁⲙⲟⲩ

Here the scribe apparently renders an arbitrary phonetic (but prosodically unjustified) close juncture between two words.

Somewhat different is 8:8 ⲁⲩⲟⲩⲁϩⲟⲩ ⲛⲛ̄ⲥⲱⲥ, where the intrusive ⲛ obviously bridges the hiatus between final vowel and vocalic onset (and thus outrules initial [ʔ]).

Haplography, under this category, is likewise found with nasals, and once involving an occlusive:

4:27 ⲉⲛⲉ (for ⲉⲛ ⲛⲉ)

9:12 ⲟⲩⲁⲛⲓⲙ (for ⲟⲩⲁⲛ ⲛⲓⲙ)
12:16 ⲧⲁⲕⲭⲱ (for (ⲛ)ⲧⲁⲕ ⲕⲭⲱ)

Glide dittography/haplography

Just as the double spelling of nasals, the gemination of glides cannot be regarded as a "dittography" in the technical sense. This gemination is probably not even to be considered "unorthographic" when it occurs inside lexical words (i.e., in those cases where the scribe follows a somewhat unusual convention for the spelling of intervocalic *w*: ⲙⲁⲩⲟⲩⲉⲧ and ⲧⲉⲩⲟⲩⲁ⸗; see below), and it does not occur at the boundary between phonological words. It is, however, found inside phonological words at morph boundaries, and this usage probably reflects the actual "glide" nature of the sounds involved, by supplying a smooth vocalic departure to the consonantal glide onset of the second element.

6:12 ⲛ̄ⲧⲁ{ⲉⲓ}ⲓⲱⲁⲛⲛⲏⲥ
7:18; 10:11 ⲛ{ⲓ}ⲓ̅ⲥ̅
12:11 ⲥⲁⲕ{ⲟⲩ}ⲟⲩⲁⲛ

(Note that the alternation between ⲉⲓ and ⲓ fully conforms to orthographic rules).

As a complementary phenomenon, glide "haplography", or the single-grapheme spelling of two adjacent glides of the same sort, is likewise found inside phonological words at morph boundaries:

4:16 ⲛⲧⲁⲉⲓ (for ⲛⲧⲁⲉⲓ-ⲉⲓ)
8:28 ⲁϥϥⲓⲉⲧϥ (for ⲁϥ-ϥⲓ-ⲉⲓⲉⲧϥ)
11:13 ⲁⲩϯⲡⲟⲩⲁⲉⲓⲉ (for ⲁⲩ-ϯ-ⲡⲟⲩ-ⲟⲩⲁⲉⲓⲉ)
12:1 ⲁⲉⲓ (for ⲁⲉⲓ-ⲉⲓ)

This convention, which is strictly followed by the scribe of T (cf. Quecke 1984b: 324f. with note 121; see also Quecke 1984a: 43, 45), appears to be broken in at least one case in the present MS (13:12, i.e. Jn 12:46). The necessary condition for this conventional simplification of glide spelling, at least in the four John MSS involved (Bohairic Bodmer III, Sahidic Pal. Rib., and the two Subachmimic MSS), seems to be that at least one of the two glides must function as a consonant. Of the four examples listed above, the sequence is C—V in the first and fourth cases, and V—C in the second and third. As Quecke has noted, this convention does not apply to V—V (for instance, ϯⲉⲓⲣⲉ). It does, however, apply to C-C (no case attested in the present MS, but T has ⲁⲉⲓⲁ- for ⲁⲉⲓ-ⲉⲓⲁ- Jn 9:11, 15 and ⲡⲉⲉⲓⲁϩⲉⲥⲁⲩ for ⲡⲉⲉⲓ-ⲉⲓⲁϩⲉ-ⲥⲁⲩ Jn 10:16).

Vowel dittography/haplography

(a) Inside a lexical word:
There is no double spelling of vowels inside lexical words except for those cases where the double vowel is conventionally used to indicate the presence of phonological /ʔ/ (in this dialect, this includes ⲱⲱϩ, ⲁⲁϩ "live"). As to haplography, "missing" double vowels are also largely conventional here (e.g., ϣⲟⲡ, ⲉ=ϥ).

(b) At morph boundary but inside a phonological word:
Dittography is extremely rare here:
8:16 ϫⲉ{ⲉ}ⲛⲧⲁⲧⲛⲕⲁⲁϥ
In this case the scribe seems to extend the orthographic convention for post-stress word-interior /ʔ/ to a pre-stress syllable (secondary stress?), signifying something like [čeʔn̩tatn̩-]. Haplography is found twice with pre-stress ⲁ:
6:7 ⲛ̄ⲥⲁ<ⲁ>ⲙⲁϩⲧⲉ
10:19 ⲉⲥⲁ<ⲁ>ⲣⲏϩ

(c) At the boundary between phonological words

(c_1) Doubling the stressed vowel of the first word:
5:17 ⲁⲗⲗⲁ{ⲁ}ⲛⲧⲱⲧⲛ
6:13 -ⲟⲩⲉⲉ{ⲉ}ⲇⲉ
10:1 ϭⲉ{ⲉ}[ⲛ]ⲥⲁⲓ̅ⲥ̅
13:14 -ⲟⲩⲉⲉ{ⲉ}ⲥⲱⲧⲙ
The triple spelling of ⲉ in ⲟⲩⲉⲉ, of course, can hardly be regarded as a legitimate doubling of the stressed vowel since the presence of /ʔ/ is already indicated by the standard spelling ⲟⲩⲉⲉ (a characteristic feature of this dialect). The uncertainty of the scribe in spelling this element may be illustrated by the complementary cases of possible haplography:
5:7 ⲛⲁⲟⲩⲉ<ⲉ> ⲉⲛ
9:14 ⲟⲩⲟ<ⲉ> ⲇⲉ
10:9 ⲛⲉⲟⲩⲉ<ⲉ> ⲡⲉ
However, the remaining two cases (5:17 and 10:1) may represent a more or less "legitimate" use of unorthographic double vowel spelling to indicate the presence of post-stress aleph in connection with a following sonorant (in these cases: [ʔn̩]). For 5:17 cf. ⲁⲗⲗⲁ{ⲁ}ⲟⲩⲁ Jn 19:34 (Quecke 1984a); for occasional double spellings of word-final stressed vowels followed by aleph cf., e.g., Quecke 1984a: 44, or the relatively broad attestation in P. Bodmer VI (Kasser 1960), where ⲁⲡⲉⲉ̅ and ⲙⲁⲁ are frequent variant spellings of ⲁⲡⲉ and ⲙⲁ, respectively, paralleled by hapax lego-

mena such as ⲙⲟⲩⲟⲩ (for ⲙⲟⲩ) and ⲛ̄ⲛⲉⲉ̄ (for ⲛ̄ⲛⲉ). In some MSS, this technique is also used to signify the aleph suffix of the 2nd fem. sing. (e.g., ϫⲡⲟⲟ̄ NHC VII 94:27).

(c_2) Doubling the unstressed final vowel of the first word (the most frequent type)

4:24 ϯⲥⲁⲩⲛⲉ{ⲉ}ⲛⲛⲱⲉⲓ
4:32 ⲙⲛ̄ⲗⲁⲩⲉ{ⲉ}ϥⲓ
5:34 ⲙⲡⲛⲟⲩⲧⲉ{ⲉ}ϣⲱⲡⲉ
6:14 ϩⲛⲃⲏⲑⲁⲛⲓⲁ{ⲁ}ⲡϯⲙⲉ
8:15 ϩⲛⲡⲉⲡ̅ⲛ̅ⲁ̅{ⲁ}ⲛⲑⲉ
9:18 ⲥⲣ̄ⲛⲁϥⲣⲉ{ⲉ}ⲛⲉⲛ
11:1 ⲧϣⲉⲉⲣⲉ{ⲉ}ⲛⲥⲉⲓⲱⲛ

Cf. also 11:13 ϩⲛⲡϣⲁⲉⲓⲉ{ⲉ ⲛ}ⲛⲉⲉⲓ.

Here the scribe apparently again extends the convention for post-stress /ʔ/ to a non-stressed syllable, where phonetic [ʔ] might or might not be expected. At any rate, this seems to be more than just a series of mistakes.

(c_3) Doubling the onset vowel of the second word

7:16 ⲁⲥⲓ{ⲁ}ⲁⲃⲁⲗ

This may be a plain mistake, since it cannot be a graphic representation of [ʔ] between ⲁⲥⲓ and ⲁⲃⲁⲗ.

To conclude, a very large number of the unusual spellings listed under three principal categories can hardly be regarded as mistakes in the narrower sense of "slips of the pen" on the part of the scribe. While some few of them may be reflective of early (non-classical) orthographic tendencies, the greater part of them would perhaps be more adequately described as "he did not know any better", although he may have tried his best. Orthography is a difficult matter, and its mastership requires a high degree of learning. Most of the spellings quoted are "errors" in the sense that, with some good reason, they would not pass as acceptable variant spellings in any known orthographical system of any literary dialect of Coptic.

One may ask whether this MS is more likely to be a relatively good copy made from a bad exemplar or a relatively bad copy made from a good exemplar (it is certainly not bad enough to be a bad copy from a bad exemplar). I would strongly favour the latter alternative. The deficiencies of this MS are almost exclusively due to awkward spellings including some omissions and confusions on the part of the copyist. Behind these deficiencies we can still discern the contours of a source text with a high degree of grammatical coherence and orthographical consistency, comparable to, but not identical with, the text extant in T.

The system of actual "orthography" underlying the text of the present MS can be briefly described with regard to the issues that are of central importance to any Coptic orthographic system: (1) the use of superlineation and the spelling of sonorant syllables, (2) the (regular) use of double vowels, (3) the rules for spelling *i/y*, and (4) the rules for spelling *u/w*.

As to superlineation, it is interesting to see that the scribe never "forgot" to draw a line when he used *nomina sacra* abbreviations: ⲓ̅ⲥ̅, ⲡⲉⲭ̅ⲥ̅ (once ⲡⲉⲭ̅ⲣ̅ⲥ̅), and ⲡⲉⲡ̅ⲛ̅ⲁ̅ (there is no such marking with proper nouns that are spelled out). Thus the scribe was well aware of the fact that the identity of these forms fully depends on the presence of superlineation. This is not equally true for the usual supralinear "stroke" that indicates syllabic unity, and the use of this stroke in the present MS is more or less irregular. It is used relatively frequently with nasals and liquids that represent a syllable of their own, such as initial ⲛ̄- or ⲙ̄-, but also ⲣ̄-, whether word-initial or otherwise. When the sonorant is preceded by a tautosyllabic consonant (e.g., in ⲙⲣ̄ⲣⲉ-, ⲡⲣ̄ⲣⲟ, ⲡⲣ̄ⲡⲉⲉ, ⲙⲡⲣ̄-, ⲥⲣ̄-; probably also in ⲁϥⲣ̄- and ⲉϥⲣ̄-), the stroke is placed over the sonorant alone (thus even in -ϣⲃⲣ̄- 7:8). In ⲙⲛ̄- (negative existential) the stroke is once found above the initial ⲙ, but usually above the ⲛ, as it is in the preposition ⲙⲛ̄-. With the 3rd masc. sing. Conjunctive preformative ⲛϥ-, the stroke is placed over ϥ (if at all). It is also used with the ϥ in ϥⲧⲁⲩ following ⲡⲉϥ- (8:25). The stroke extending over more than one letter is of extreme rarity: ⲡϣ̅ⲥ̅ 4:17; ϩⲏⲧ̅ⲥ̅ 6:34; ⲟⲩⲁⲉⲧ̅ⲥ̅ 11:22 (but ⲉⲛⲛⲁϣϥ̄ 11:23); ϣⲧⲁⲣⲧ̅ⲣ̅ 11:30. There seems to be some regularity in the absence of the stroke with entire sets of forms; thus, for example, the stroke is never used in prenominal forms of prepositions consisting of more than one letter (such as ϩⲛ-, ϩⲓⲧⲛ-, ⲁϫⲛ-), with the negative ⲧⲙ-, or with 2nd plural indicators (as in ⲡⲉⲧⲛ-, ⲉⲧⲉⲧⲛ-; ⲛ̄ⲧⲱⲧⲛ, ⲛⲏⲧⲛ, etc.). In the presuffixal form of the preposition ⲙⲛ̄- (ⲛⲙⲙⲉ⸗), the stroke is usually not used (except once in ⲛ̄ⲙⲏⲓ 14:17). The rule may have been for the scribe to use the stroke with nasals generally when one letter stands for a syllable; in two-letter syllables he used it, however inconsistently, with "real" consonants and with liquids (ⲣ), but not with nasals (except for the sequence ⲙⲛ-).

Some of the occasions for stroke usage in other MSS do not apply here, since consonant clusters are often vocalized. For pre-stress non-nasal sonorant syllables, we find the spellings ⲃⲓⲗⲃⲓⲗⲉ 11:20 and ⲥⲉⲗⲥⲱⲗⲟⲩ 7:14 (cf. ⲥⲉⲃⲧⲱⲧ, but ⲃⲗⲃⲓⲗⲉ and ⲥⲗⲥⲱⲗⲟⲩ in T;

with non-sonorant: ⲥⲉⲡⲥⲱⲡ in both MSS). In ϩⲛⲙⲁⲥ 7:17 the first syllable does not receive a vowel because the sonorant is a nasal and is itself intrusive. The Relative Present varies between ⲉⲧⲕ-, ⲉⲧϥ-, ⲉⲧⲥ- and ⲉⲧⲉⲕ-, ⲉⲧⲉϥ-. The typically Subachmimic /a/ vocalization of construct infinitives (cf. ⲗⲁϭ-, ⲛⲁϫ-, ⲧⲁⲛϩⲁ-, ⲟⲩⲁⲛⲉϩ-, ⲟⲩⲁϣⲃ-, etc., in T) is here attested by ⲧⲁⲃϩ- 7:20 (as in ManiPs, but ⲧ̄ⲃ̄ϩ̄- T) and ⲧⲁϩⲁⲥ- 6:15 (ⲧⲁϩⲥ- T). Another type of vocalic resolution, that of word-final /RC/ (sonorant-consonant) clusters, is also rather frequent here. Such clusters are spelled without vowel only in ⲟⲩⲱⲛϣ and ϭⲱⲗⲡ, and once in ⲧⲏⲣϥ (also in ⲥⲙⲁⲙⲁⲛⲧ, but here the sonorant is itself intrusive: post-aleph, cf. ⲥⲙⲁⲙⲁⲁⲛⲧ in T). Vocalic resolution represented by ⲉ is found in ⲃⲁⲗⲉϥ, ⲧⲏⲣⲉϥ (once), ⲱⲃⲉϣ and ϣⲁⲣⲉⲡ; represented by ⲁ in ⲧⲱⲙⲁⲧ. (Note that there is a similar, but weaker, tendency in T, which spells ⲃⲁⲗϥ, ⲧⲱⲙ̄ⲧ, ⲧⲏⲣϥ [except ⲧⲏⲣⲉⲕ Jn 9:34], ⲱⲃϣ, but ⲙⲁⲣⲉϥ [Jn 18:12], and always ⲛ̄ϣⲁⲣⲉⲡ and ϫⲓⲛϣⲁⲣⲉⲡ [but once ⲣ̄ϣⲁⲣⲡ Jn 5:7]; cf. also occasional ⲟⲩⲱⲛⲉϩ Jn 17:26). While this habit usually applies only to non-glide sonorants (with word-final /GC/ being usually resolved into the respective [VC] sequence: ⲟⲩⲁⲉⲓϣ, ⲣⲁⲟⲩϣ, -ⲥⲛⲁⲟⲩⲥ, etc.), in the present MS it is extended in one case to word-final /wC/: the stative of the verb ⲙⲟⲩⲟⲩⲧ (whose infinitive is here only attested with construct forms: ⲙⲁⲟⲩⲧ= and ⲙⲁⲟⲩⲧ-) is spelled ⲙⲁⲩⲟⲩⲉⲧ instead of "normal" ⲙⲁⲟⲩⲧ (thus T). For /CNC/ clusters, which are not attested, see the note on Jn 12:47.

The use of double vowel spelling at those occasions where it is usually expected to occur (i.e., for non-final, post-stress aleph) appears to be fairly well normalized. This MS agrees with T (and with most other Subachmimic MSS) in spelling double vowels in words such as ⲕⲁⲁ=, ⲕⲉⲉⲥⲉ, ⲙϩⲉⲉⲩ, (ⲛ̄)ⲧⲟⲟⲧ=, and ϣⲉⲉⲣⲉ, but single vowels in ⲉⲧⲃⲏⲧ=, ⲗⲁⲩⲉ, ⲙⲏϣⲉ, ⲙⲁⲛⲉ, ⲥⲁⲛⲉ, ϣⲱⲛⲉ. It also largely shares with T the peculiar word-final double vowel spelling in ⲟⲩⲉⲉ "one" and ⲣ̄ⲡⲉⲉ "temple", and it has ⲱⲱϩ, ⲁⲁϩ where T spells ⲱⲱⲛϩ, ⲁⲁⲛϩ. But it is distinct from T (and closer to Manichaean MSS) in spelling consistently ⲉ=, ⲧⲉ= (instead of ⲉⲉ=, ⲧⲉⲉ=) for the status pronominales of the verbs ϯ and ⲉⲓⲣⲉ, as well as ϣⲟⲡ (ϣⲟⲟⲡ T) and ⲙⲁϩⲉ (ⲙⲁⲁϩⲉ T), and it does not share the apparent consistency of T in the peculiar double vowel spelling for the st. pronom. of the preposition ϣⲁ-: ϣⲁⲁⲣⲁ= (ϣⲁⲁⲣⲁϥ 6:10; 8:4; but ϣⲁⲣⲁϥ 6:17; 7:7; lac. 5:34; 12:12). With this exception, and with the exception of ⲟⲩⲉⲉ, which occasionally suffers from haplography or dittography (see above), the spelling of simple or double vowels is consistent for each of these forms.

Apart from double vowel spelling, the letter ⲉ is used word-finally after stressed vowels in a number of lexemes in Achmimic-Subachmimic MSS. Some of the most common examples are ⲛⲁⲉ *A L* "have mercy"; ⲙⲓⲉ A, ⲙⲏⲉ *L4 L6* "truth"; ⲕⲱⲉ *L6* (NH and AP) "put". The interpretation of this phenomenon may be left aside here (for a brief indication of the issue see Quecke 1984b: 289 note 1), although it seems fairly plausible to regard this as an indication of final aleph. This group of regular spellings is represented in the present MS by three lexemes: ⲃⲁⲉ "branch of date-palm" (= T and *A L* in general), ⲙⲓⲉ "truth" (T ⲙⲏⲉ), and ϥⲟⲩⲉ "hair" (T ϥⲱⲉ). In these three lexemes, final ⲉ is found regularly in the orthography of all Achmimic and Subachmimic dialects.

As to the graphic representation of <i/y>, the most salient feature of the present MS is the absence of any diaeresis. Thus, while the number of allographs available in Upper Egyptian MSS (including T) is usually three: ⲉⲓ, ⲓ̈, and ⲓ (sometimes also ⲉ͡ⲓ, but this is probably a subvariant of ⲉⲓ), in this MS this number is reduced to two: ⲉⲓ and ⲓ. The distribution of these allographs is largely governed by the simple principle that ⲉⲓ is used word-initially and after vowels (except ⲏ), whereas ⲓ is used after consonants (and either sign may be used after ⲏ). This principle, which likewise governs the usage in T, hardly comes as a surprise in a non-Manichaean Subachmimic MS. Deviations from this principle, here, occur mainly with the 1st sing. pronoun suffixed to vocalic conjugation bases, but they are extremely rare in forms with tautomorphic i/y. Thus we have ⲙⲁⲉⲓⲉ, ⲟⲩⲁⲉⲓⲉ, ϣⲁⲉⲓⲉ, ϩⲁⲉⲓⲉ (an exception is ⲭⲁⲓⲉ 9:29, pressed at the end of a line; T has ϫⲁⲉⲓⲉ) ⲁⲉⲓⲕ, ⲙⲁⲉⲓⲛ, ⲧⲟⲩⲃⲁⲉⲓⲧ, ⲟⲩⲁⲉⲓⲛ, ⲭⲁⲉⲓⲥ; ⲟⲩⲁⲉⲓⲛⲓⲛ (sic), ϩⲁⲉⲓⲛⲉ; ⲥⲧⲁⲉⲓ, ⲭⲁⲉⲓ-, ϭⲃⲁⲉⲓ; ⲡⲉⲉⲓ etc., ⲡⲉⲉⲓ- etc., ⲧⲉⲉⲓⲧⲟⲩ, ⲟⲩⲭⲉⲉⲓ; ⲕⲟⲩⲉⲓ; and, on the other hand, ⲃⲓⲗⲃⲓⲗⲉ, ⲣⲓⲙⲉ, ϣⲓⲛⲉ, ⲭⲓⲟⲩⲉ, ϭⲓⲛⲉ; ⲛⲓⲙ, ϭⲓⲭ; ⲙⲓⲉ, ϥⲓ; ⲁⲣⲓ-, (ⲛ̄)ϭⲓ-, ϥⲓ-, ϩⲓ-, ⲭⲓ-/ⲭⲓⲧ⸗. An exception is ⲭⲉⲓ-, the *i*-vocalized variant of ⲭⲉ- (st. nom. of ⲭⲱ) used once in the compound ⲭⲉⲓⲟⲩⲁ "blaspheme" — which would usually be spelled ϫⲓⲟⲩⲁ (T has ϫⲉⲩⲁ). As far as preceding ⲏ is concerned, we find the same distribution as in T: ⲏⲉⲓ "house" but ϩⲣⲏⲓ (except that T has the diaeresis: ϩⲣⲏⲓ̈).

The same principle applies to the suffix of 1st sing. in word-final position [ⲁⲣⲁⲉⲓ, (ⲙ)ⲙⲁⲉⲓ, ⲧⲉⲩⲟⲩⲁⲉⲓ, ϩⲁⲣⲁⲉⲓ; ⲛⲱⲉⲓ, (ⲛ̄)ⲥⲱⲉⲓ]. After suffix-conditioned ⲏ, where T spells -ⲉⲓ, the present MS seems to prefer -ι (νηι twice, ν̄μηι, ογνтηι), but once has νηєι (5:22). In conjugation forms, however, non-use of the diaeresis creates graphic ambiguity if the basic vowel is є. But even T here sometimes violates the principle (cf. the exceptional spellings єινα- Jn 4:34; 11:11; єια- 5:30; 6:39; see also Quecke 1984b: 293). In the present MS we find єєι- (the normal spelling in T) only once (for the Second Pres.), besides єινα- (three times), єιтм- (once), and νєι- (once), but for the Perfect conjugations the good distribution between αєι- and мπι- without exception (the Temporal тαρєι- 8:34 is a mistake for тαρєϥ-, otherwise it should of course be тαρι-).

What is less common, although by now well attested in several early MSS, is the extension of the same principles to those cases where lexeme-initial єι is combined with preceding elements (in close junction: conjugation and determination). In this regard, the present MS is at least as rigorous as T. While it always spells єι when such forms are used independently (here only #єιc- and #єι; #єιρє is not attested) or after vowels (νcє-єι 9:13, ϯєιρє passim, α-єια- 14:15, єтνα-єια- 14:12), it regularly has ι after consonants: єκιρє and νєϥιρє; αϥι, πατєϥι, ν̄ταρєϥι, νϥι and even μαϥι 4:14 (where T has μαϥєι), as well as αcι and νταρєcι. (Note that, although T regularly spells єϥιρє, єтϥιρє, αтρєϥιρє, etc., αϥινє and єϥινє, there is an exceptional єϥєιογє in Jn 13:12; but even with the verb єιογє/єια-, where the initial *i* is originally "consonantal", T usually spells κιογє, αϥια-, ν̄κια- and ν̄ϥια-). Postvocalic *u* and syllabic sonorants, for this issue, count as vowels since they cannot become tautosyllabic with following *i*; thus we have αγ-єι and єιтм-єι. The same rule, in principle, applies to nouns with initial єι. While the full form is spelled in each case for єιωт, єιαϩєcαγ, and єιω when these forms follow πα-, ογ-, or ν-, their initial *i/y* is represented by ι when following π-, with which it forms a syllable: πιωт 6:1; πιαϩє-cαγ 4:27 (for this noun see *supra*, p. 73). (Scribes who follow this principle may even find no problem in extending it to the verb єι, cf. the spelling πι NHC VI 19:13; VIII 9:6). The present excerpt of the gospel, however, does not provide a higher number of occasions for this combination to occur, so that we cannot see whether this MS would share the inconsistency of T with regard to πιωт and πєιωт, or whether it would take a more consistent stance.

The spelling of *i/y* in lexemes of Greek origin primarily follows Greek, not Coptic, conventions. Deviations from Greek standards usually should be in agreement with the rules for Coptic words (this is the case, for instance, in διακονι, λοϭι, and χρια).

As to the graphic representation of <u/w>, the usual number of allographs available in most Coptic dialects is two: ⲟⲩ and ⲩ. In the present MS their number is three, adding ⲩⲟⲩ. Apart from this particularity, the rules for <u/w> are the same as in T. They can be described as follows.

With ⲟⲩ being the only admissible allograph word-initially, after consonants and sonorants (whether syllabic or not), after (ⲉ)ⲓ (e.g., ϫⲓⲟⲩⲉ), after (ⲟ)ⲩ (e.g., ϣⲟⲩⲟⲩⲧ) and after ⲟ (e.g., ϩⲟⲟⲩ) in Coptic in general, it is merely the spelling of the grapheme when following ⲁ, ⲉ, ⲏ, and ⲱ, and in combinations with prefixed elements, that can be at issue here. ⲱ usually joins ⲟ in demanding ⲟⲩ (here only in ϩⲱⲟⲩⲛ and ϩⲱⲟⲩⲧⲛ, as in T). For the spelling of <u/w> after <a, e, ē> the following principles may be established (although the attestation is not always broad enough to support the formulation of a "rule"). ⲟⲩ is only spelled in one particular case: before a consonant (or sonorant) when this consonant is, or has a chance of being, tautosyllabic (so that it is relatively easy to draw a syllable boundary between *u/w* and the preceding vowel). Such tautosyllabic consonants are found, for instance, in ⲣⲁⲟⲩϣ, -ⲥⲛⲁⲟⲩⲥ, ⲙⲁⲟⲩⲧ-, and ⲙⲁⲟⲩⲧϥ in both MSS (the last-mentioned verb is not attested with a suffix of 1st or 3rd plural in either MS; if the condition specified here is correct, one would expect ⲙⲁⲩⲧ̄ⲛ̄ and ⲙⲁⲩⲧⲟⲩ). The opposite situation is found in those cases where the consonant following u/w has no chance of being tautosyllabic since it is followed by a vowel: ⲥⲁⲩⲛⲉ, ⲥⲁⲩϩⲟⲩ. The tautosyllabic consonant can also represent a suffix: ⲧⲛⲛⲁⲟⲩϥ (but ⲧⲛⲛⲁⲩ word-finally, only attested in T).

NB: Stem-final /w/ in the verb ⲧⲛⲛⲁⲩ is "consonantal" not only in the phonetic representation of the *status absolutus* form but also in terms of morphophonology (a "radical"); thus the verb requires the suffix /-t/ as allomorph for the 1st sing. Once it takes the /-t/, however, the stem-final /w/ forms a syllable with it (as it does with any other uniconsonantal suffix) and thus becomes vocalic: ⲧⲛⲛⲁⲟⲩⲧ (attested only in T).

In all other cases where it follows <a, e>, the grapheme <u/w> is spelled ⲩ: word-finally as in ⲉⲥⲁⲩ, ⲛ̄ⲧⲁⲩ, ⲙⲁⲩ, etc., (ⲙ̄)ⲙⲉⲩ, ⲙ̄ϩⲉⲉⲩ, ⲛⲉⲩ, ⲥⲛⲉⲩ (this also applies to the suffix in prefixed elements like ⲁⲩ- and ⲉⲩ-), preceding single vowels, whether stressed (ⲁⲩⲱ, ⲧⲉⲩⲁϥ) or unstressed (ⲕⲁⲩⲉ, ⲗⲁⲩⲉ, ⲙⲉⲩⲉ), or preceding clearly heterosyllabic consonants (ⲥⲁⲩⲛⲉ, ⲥⲁⲩϩⲟⲩ). The same principles may hold for the spelling of *u/w* after ⲏ, but this is attested in the present

MS only word-finally (ⲉⲣⲏⲩ, (ⲛ)ⲛⲏⲩ, ϩⲏⲩ) and intervocalically (ϩⲃⲏⲩⲉ).

As a subvariant of intervocalic ⲩ for <u/w> we find a sort of double-glide spelling: ⲩⲟⲩ, a very peculiar, but apparantly regular feature of our MS. It seems to be applicable only when the vowel that follows <u/w> is "checked" (thus not in ⲕⲁⲩⲉ, ⲗⲁⲩⲉ, ⲙⲉⲩⲉ here; but cf. ⲙⲉⲩⲟⲩⲉ as a variant spelling, e.g., in NHC I 85:24f,; Mani-Keph 134:1; Act 14:19 Hintze-Schenke). Nevertheless, the two forms for which it is actually attested here (ⲙⲁⲩⲟⲩⲉⲧ and ⲧⲉⲩⲟⲩⲁⲉⲓ) are very different in nature both morphologically and phonologically. The stative of ⲙⲟⲩⲟⲩⲧ, on the one hand, for which T has the regular ⲙⲁⲟⲩⲧ, is here subject to the optional *e*-vocalization of the final cluster in /CV́RC/, here extended to /CV́GC/ (see above). For some reason, it seems, a spelling *ⲙⲁⲩⲉⲧ was not acceptable to the scribe; but post-stress ⲩⲟⲩ is generally very rare (one other example is ⲗⲁⲩⲟⲩ Act 4:17 Hintze-Schenke). In the causative verb ⲧⲉⲩⲟ (ⲧⲉⲩⲁ⸗, ⲧⲉⲩⲟⲩⲁ⸗), on the other hand, it is the stressed vowel that follows /w/. This is in good agreement with most other forms in which ⲩⲟⲩ occasionally occurs (cf. in particular ⲧⲉⲩⲟⲩⲁϥ NHC II 54:5; and see Hintze-Schenke 1970:17f.). It is hard to see, however, why this form should be treated differently for the suffixes of 1st sing. (ⲧⲉⲩⲟⲩⲁⲉⲓ) and 3rd masc. sing. (ⲧⲉⲩⲁϥ, only one occurrence); T has ⲧⲉⲩⲁⲉⲓ and ⲧⲉⲩⲁϥ.

As to the status of ⲩⲟⲩ in the system of orthography, two different aspects should be kept apart. In relation to the single-representation allographs ⲟⲩ and ⲩ, the compound allograph ⲩⲟⲩ is a "geminating" variant of ⲩ in its intervocalic usage (i.e., at syllable boundaries). At the same time, and from the point of view of "glide gemination", however, ⲩⲟⲩ is the postvocalic allograph of ⲟⲩⲟⲩ, which would be used, if at all, word-initially and after consonants and sonorants (for examples see Hintze-Schenke 1970:18f.).

The suffix of 3rd plural {-w} is generally treated like a lexeme-internal element and thus obeys the same rules: ⲁⲣⲁⲩ, ⲙⲙⲁⲩ, etc., ⲡⲁϫⲉⲩ, but ⲧⲉⲧⲟⲩ, ⲧⲏⲣⲟⲩ, ⲟⲩⲁϩⲟⲩ, etc.; thus also in suffixed conjugation bases: ⲉⲩ- and ⲁⲩ-. Initial ⲟⲩ that is subject to prefixation by vocalic elements is here mainly represented by the singular form of the indefinite article and by the existential predicate ⲟⲩⲛ-. The spelling of these is exactly the same as in T: it is ⲩ with all single-letter prefixes and the bare converters (thus, e.g., with preposition ⲁ-ⲩ-; Perfect ⲁ-ⲩ-; Preterite usually ⲛⲉ-ⲩ- 8:22; 10:16 [but cf. ⲛⲉⲟⲩⲉ 10:9, rarely ⲛⲉ-ⲟⲩ-in T] and ⲛⲉ-ⲩⲛ-, Circumst. ⲉ-ⲩⲛ-), but ⲟⲩ with all prenominal conjugation forms consisting of more than one letter (thus ⲉⲣⲉ-ⲟⲩ- 9:19 [cf. also ⲉⲣⲉⲟⲩⲉⲉ 10:4] and ⲛⲉ-ⲟⲩ- Neg. Fut. 16:25; ⲛⲉⲣⲉ-ⲟⲩ- and ⲛ̄ⲧⲁⲣⲉ-ⲟⲩ- are not attested).

As in T, and in the Achmimic and the Manichaean MSS, the special allomorph of the definite article preceding two subsequent consonants (ⲡⲉ- etc.) is not used in the present MS with ordinary lexemes. In contrast to those other MSS, however, it is used in the *nomina sacra* ⲡⲉⲭ̅ⲥ̅ (once ⲡⲉⲭ̅ⲣ̅ⲥ̅) and ⲡⲉⲡ̅ⲛ̅ⲁ̅. An exception would be 4:18 ⲧⲉⲯⲩⲭⲏ, but this is very probably a scribal error (for ⲧⲉϥⲯⲩⲭⲏ).

3. The Dialect

Before discussing at large the dialect affliliation of Chester Beatty ac. 1390, it may be appropriate to mention three of the more unusual phenomena, or rarities, that this MS adds to our knowledge of the lexical and morphological inventory of Coptic: (1) the word ⲱϣ "thing", (2) the form ⲉⲣⲉⲧⲉⲧⲛ(ⲛⲁ)- (alongside the normal form ⲉⲧⲉⲧⲛ(ⲛⲁ)-) for the Second Present (Future) with the suffix of 2nd plural, and (3) the apparently unprecedented use of the conjunction ⲉϣⲱⲡⲉ. For the second of these items see below, variable no. 32.

The word ⲱϣ occurs in John 10:33 ⲉ̣[ⲛⲁ]ⲛⲁⲭⲱⲛⲉ ⲉⲛ ⲁⲣⲁⲕ ⲉⲧⲃⲉⲱϣ ⲉⲛⲁⲛⲟⲩϥ "it is not for a good thing that we stone you". The use of ⲱϣ, instead of ϩⲱⲃ (thus T along with sa), is somewhat surprising in this context, since the sentence quoted is supposed to be a refusal of the claim implicit in Jesus' question immediately preceding it, where the present MS, too, uses ϩⲱⲃ (i.e., ϩ[ⲁϩ] ⲛϩⲱⲃ ⲉⲛⲁⲛⲟⲩϥ and ⲉⲧⲃⲉⲉϣ ⲙⲙ[ⲁⲩ] ⲛ̄ϩⲱⲃ). But there can be hardly any doubt about the identity of ⲱϣ "thing" in the response: it is one of those words current at earlier stages of Egyptian (*jḫ.t*) that Egyptologists have been expecting to reappear in Coptic for quite some time. Vycichl (1983:20a, s.v. ⲁϣ) gives **ōẖe* as the expected Coptic (Achmimic) form; regarding the present occurrence of the word he proposes to assume haplography of the epsilon, i.e., to read ⲉⲧⲃⲉⲱϣⲉ <ⲉ>ⲛⲁⲛⲟⲩϥ (personal letter, June 2, 1987). This does not seem advisable, however, in what is the first attestation of a word in Coptic — and it may be unnecessary. Jürgen Osing (personal letter, March 31, 1988) confirms that the form **ōẖ* (i.e., ⲱϩ *A*, ⲱϣ *S L*) had always been considered likewise probable: the late vocalization of the word could so far only be inferred from the Greek transcription Ασυχις of the Egyptian name *ʿšꜣ-jẖt*, and this did not allow precise predictions about the termination of the form (presence or absence of a final vowel). Vycichl (1983:20a) considers this word to be the direct ancestor of the interrogative pronoun *jḫ* (Coptic ⲁϣ *S*, ⲉϣ *L* etc.), but Osing points out that direct derivation of one form from the other is impossible in either direction, the precise nature of their etymological relationship thus being unclear.

It remains to be seen whether the attested survival of ⲱϣ "thing" in an Upper Egyptian dialect of Coptic can shed new light on any of those passages in early texts where the interpretation of some form (ⲟⲩ)ⲱϣ on the basis of our previous knowledge met with problems.

As to the third of the above-mentioned items, this MS is unique in the lack of distinction between the conjunction ⲉϣⲱⲡⲉ (*S F L*; ⲉϣⲱⲡ *B*, ⲉϣⲟⲡⲉ *M*, ⲉϩⲱⲡⲉ *A*) "if it happens that" and an equivalent of Sahidic ⲉϣϫⲉ- "if it is true that" (ⲉϣⲡⲉ- in T; ϩⲡⲉ-/ⲉⲓϩⲡⲉ- *A*, ϣⲡⲉ-/ⲉⲓϣⲡⲉ- *L6*, etc.; see Funk 1985a:23f.). The form ⲉϣⲱⲡⲉ is used throughout these fragments whenever a neutral ("non-remote") conditional clause is introduced by means of a conjunction, although this form should be expected to occur only in the two cases of an "open" (or "non-manifest") condition, while an equivalent of ⲉϣϫⲉ- would be expected to introduce the protasis of a "closed" (or "manifest") condition, which occurs five times in this MS (for references see the Indices; for the distinction between "open" and "closed" conditions see Funk 1984:174). This fact seems open to various interpretations, which may be of some interest also in the light of Černy's explanation of the form ⲉϣⲡⲉ (found exclusively in T) as a "contraction" of ⲉϣⲱⲡⲉ (Černy 1976:41). Such interpretations include the following:

(1) The distinction between open and closed conditionals was lost in the *état de langue* of which this MS is a text specimen. (This may or may not be connected with Černy's explanation of ⲉϣⲡⲉ).

(2) The distinction between the two types of conditionals was maintained in the *état de langue* but not overtly expressed in orthography. In this case the text would actually have two forms spelled ⲉϣⲱⲡⲉ, which are formally distinct only in terms of prosody (which is perhaps the most important formal feature of the distinction between the two conjunctions in all Upper Egyptian dialects except *L4* and *S*): a prosodically autonomous, unrestricted conjunction ⲉϣⲱⲡⲉ "if it happens that", which may be directly followed by a second-position particle and can be used optionally to precede a Conditional conjugation form, and a proclitic, converter-like element ⲉϣⲱⲡⲉ- "if it is true that", which must be directly linked to a main structure (nucleus). This

interpretation could be verified or falsified only by observing the use of a second-position particle in the latter type of construction — which, unfortunately, does not occur in these fragments (for T, see John 5:47; 10:38; 18:8).

(3) The copyist of the present MS found ⲉϣⲡⲉ in his exemplar (in those cases where it applies) and replaced it deliberately by ⲉϣⲱⲡⲉ, perhaps because the former spelling appeared strange to him and/or he was not aware of the distinction.

(4) The copyist found ⲉϣⲡⲉ in his exemplar and replaced it inadvertently by ⲉϣⲱⲡⲉ (compare his generally unstable orthographic behaviour in several cases involving the letters ⲱ and ϣ, see above).

Of these, solution (4) is the least probable, given the scribe's consistency in spelling ⲉϣⲱⲡⲉ, solution (3) seems methodologically rather dubious, and solution (1) would be awkward for what is evidently a general typological feature of Coptic. I would suggest to take resort to either solution (1) or (3) only if solution (2) can indeed be falsified on the basis of other MSS with a similar usage. Otherwise (2) would seem to be the most natural solution: compare the prosodically similar case of the ambivalent spelling ⲛ̄ⲧⲱⲧⲛ (besides ⲁⲛⲁⲕ and ⲁⲛⲁⲛ), which may represent #ⲛ̄ⲧⲱⲧⲛ# (the "augens" and predicative pronoun) or #ⲛ̄ⲧⲱⲧⲛ- (the subject pronoun of an interlocutive nominal sentence, e.g., John 10:26, 34, which in some other orthographies is spelled ⲛ̄ⲧⲉⲧⲛ-).

At least, we have to take into consideration the possibility that the apparent identity of the two conditional conjunctions in the present MS is just a matter of graphic representation (two homographs). In this light, even Černy's "contraction" theory may gain some plausibility.

The issue of dialect affliliation of the present MS has some bearing on the classification of "Subachmimic" dialects, and thus on Coptic dialectology in general. In recent attempts at such a classification Thompson's John stood apart from the other Subachmimic MSS in several respects and was therefore found to be the sole extant witness of a distinct variety of Coptic (Funk 1985b: 135f.). This variety has been termed *L5*, and two MSS, as yet unpublished, have been taken into consideration for membership in the group of texts that may represent the same dialect: the Act of Paul MS in the Bodmer collection (cf. Kasser 1984:273f.) and the MS published here.

With the appearance of another candidate for this variety, the question has to be answered to what extent this MS agrees with the provisionally established charateristics of *L5* (or any other related dialect). Does it share the distinctness of the dialect of T from both *L4* and *L6*, and thus confirm the existence of a dialect "norm" *L5*, and does it permit us to establish a definition of *L5* that will be more than just the record of usage of a single MS?

In the discussion of dialectal features I shall stick to the "set of variables" used as a guideline in Funk 1985b, to facilitate comparison with the other Achmimic/Subachmimic manuscripts. Explicit reference to passages (in this section: chapter and verse of the gospel) will be made only in some particular cases; in general the reader is referred to the Indices.

(1) The short stressed front vowel preceding /y/ in certain groups of words is the same as in *L5 L6*, different from *A L4*. Attested are ⲙⲁⲉⲓⲛ "sign"; ⲙⲁⲉⲓⲉ "love" and ϩⲁⲉⲓⲉ "fall'; cf. also ⲟⲩⲁⲉⲓⲉ "rush" (*S* ⲟⲩⲟⲉⲓ), ϣⲁⲉⲓⲉ "festival" (*S* ϣⲁ), ⲭⲁⲓⲉ 'desert' (*S* ϫⲁⲓⲉ), ϭⲃⲁⲉⲓ "arm" (*S* ϭⲃⲟ(ⲉ)ⲓ).

(2) The short stressed front vowel preceding /w#/ in certain forms where this combination is subject to monophthongization (/ɛw/ → [ɔ:]) in *A L4*, is always maintained as /ɛ/, as in *L5 L6* (and sporadically in *L4*): ⲉⲩ "what?", (ⲙ)ⲙⲉⲩ "there", ⲛⲉⲩ "see", ⲥⲛⲉⲩ "two".

(3-5) The long stressed vowels of the front and back series conditioned by following /ʔ/ (or in word final position) require specific attention in distinguishing Subachmimic dialects. While both *A* and *L6* follow clear-cut patterns in representing these vowels, there is some fluctuation in *L5* and, to a smaller extent, *L4*. There is not sufficient attestation of sample words in the two Johns to establish a general rule, but grosso modo the Chester Beatty MS seems to confirm the indecision (or, partly, lexical division) found as a trend in T. If one counts only relatively safe cases (i.e., only words attested more than once and without variation), the picture is straightforward enough: ⲙⲏϣⲉ, ⲥⲙⲏ and ⲭⲱ (i.e. ϫⲱ), without any *A*-type vocalization. The next candidates with a more-than-once attestation, however, are ϥⲟⲩⲉ (apparently twice with this vowel, although in one case not spelled correctly) and ⲕⲟⲩ (3) besides ⲕⲱ (1). But the rest of these forms, each attested only once in the present MS, display an obvious trend when taken together: ⲟⲩϣⲓ and ⲙⲓⲉ (for e → i/_ʔ); ϣⲟⲩⲟⲩⲧ and ϭⲟⲩ (for o → u/_ʔ); all except ϭⲟⲩ differing from T. This trend indicates a closer relationship to *A* on the part of the Chester Beatty MS, compared with T, and it confirms the independence of both MSS from the particular decisions made in dialect *L6*.

This can be illustrated as follows:

A	*L5* Ch.B. 1390	*L5* T	*L6*
ⲙⲓⲉⲓϣⲉ	ⲙⲏϣⲉ (14)	ⲙⲏϣⲉ (passim)	ⲙⲏ(ⲏ)ϣⲉ
ⲥⲙⲓ	ⲥⲙⲏ (4)	ⲥⲙⲏ (6)	ⲥⲙⲏ
ⲟⲩϩⲓ	ⲟⲩϣⲓ (1)	ⲟⲩϣⲏ (4)	ⲟⲩϣⲏ
ⲙⲓⲉ	ⲙⲓⲉ (1)	ⲙⲏⲉ (passim)	ⲙⲏⲉ
ϣⲟⲩⲟⲩⲧ	ϣⲟⲩⲟⲩⲧ (1)	ϣⲱⲱⲧ (1)	ϣⲱⲱⲧ
ϥⲟⲩⲉ	ϥⲟⲩⲉ (2)	ϥⲱⲉ (2)	–
ⲥⲃⲟⲩ	—	ⲥⲃⲟⲩ (4) /ⲥⲃⲱ (4)	ⲥⲃⲱ
ϭⲟⲩ	ϭⲟⲩ (1)	ϭⲟⲩ (9) /ϭⲱ (10)	ϭⲱ
ⲕⲟⲩ	ⲕⲟⲩ (3)/ⲕⲱ (1)	ⲕⲟⲩ (9) /ⲕⲱ (5)	ⲕⲱⲉ
ϩⲃ̄ⲥⲟⲩ	—	ϩⲃ̄ⲥⲟⲩ (1)	ϩⲃ̄ⲥⲟⲩ
ϫⲟⲩ	ϫⲱ (18)	ϫⲱ (pass.) /ϫⲟⲩ (2)	ϫⲟⲩ
ⲥⲟⲩ	—	ⲥⲱ (10) /ⲥⲟⲩ (1)	ⲥⲟⲩ
ⲉⲓⲟⲩ	ⲉⲓⲱ (1)	ⲉⲓⲱ (2)	—

With the regular ("semi-Achmimicized") spelling ϩⲱⲟⲩ⸗ for the "augens" meaning "too, -self" (*S* ϩⲱⲱ⸗) the Chester Beatty MS shares one of the rare peculiarities of the dialect norm of *L5* as displayed by T.

The form ⲙⲓⲉ is also found in the third witness of *L5*, the Bodmer Act of Paul (Kasser 1984:283).

(6) There is no nasalization (as there is in *A*) of the long stressed back vowel in words like ⲙⲟⲩⲧⲉ and ⲛⲟⲩⲧⲉ.

(7) Edel's rule, which fully applies to *L6 I I7*, does not apply to the present MS: it spells ⲕⲉⲕⲉ "darkness" (twice) and ⲗⲁⲃⲉ "be mad". — The masculine form of the numeral "one" (*S* ⲟⲩⲁ), which is not subject to Edel's rule but usually conforms to that isogloss in terms of dialect usage, is [wɛʔ] as in *A L4 L5* generally, which is here spelled ⲟⲩⲉⲉ (8 times; = T) or ⲟⲩⲉ (twice). In one instance (10:9) we find the spelling that represents the standard *L6* form ([wɛy], spelled ⲟⲩⲉⲉⲓ), which may be a simple mistake.

(8-9) There are no *A*-type final sonorant syllables: ⲥⲱⲧⲙ, ⲟⲩⲱϣⲃ; ⲛⲏⲧⲛ, ⲛ(ⲙ)ⲙⲏⲧⲛ, etc.

(10) There are no *A*-type or *A*/*I7*-type laryngals: ⲟⲩϣⲓ, ⲟⲩⲱϣ, ϣ-,

ϣⲁ-, ϣⲃⲏⲣ, ϣⲏⲙ, ϣⲱⲡⲉ, ϣⲁⲣⲉⲡ, ϣⲧⲁⲣⲧⲣ, etc., on the one hand, on the other ϩⲛ-/ⲛ̄ϩⲏⲧ⸗, etc.

(11) Initial sibilant assimilation marks one of the most reliable distinctive features among Subachmimic dialects (with assimilation found in *A L6*, non-assimilation in *L4 L5*). The present MS clearly joins the non-assimilating group: ⲥⲉⲭⲉ (i.e., ⲥⲉϫⲉ), ⲥⲁⲭⲛⲉ (i.e., ⲥⲁϫⲛⲉ).

(12) The labial fricative in initial position is represented, as usually, by the letter ϥ (ϥⲓ, ϥⲟⲩⲉ, ϥⲱⲧⲉ) and thus probably to be interpreted as a labio-dental voiceless fricative, in contrast to dialect *L4* (ⲃⲓ, etc.). There is no orthographic confusion between the letters ⲃ and ϥ in various other positions (as there is in the Subachmimic first half of Nag Hammadi Codex XI).

(13-14) The stative (qualitative) of ⲉⲓⲣⲉ is not attested in this MS.

(15) As to the status pronominalis of ⲉⲓⲣⲉ and ϯ to be used with consonantal suffixes, the only form attested is ⲉ⸗ϥ (3 times), which fits into the general pattern according to which these forms, in dialect *L5* (just as in *L4 A M S*), have no /y/ after the stressed vowel (as they do in *L6 P F B*).

(16) The forms of the same verbs to be used with the suffix of the third person plural display some variation. While they always have /t/ between the stem and the suffix, this may or may not be preceded by /y/: ⲧⲉⲉⲓⲧⲟⲩ (2), but ⲉⲧⲟⲩ (3), ⲧⲉⲧⲟⲩ (1). A different kind of variation is found both in T (ⲧⲉⲉⲩ along with ⲧⲉⲉⲧⲟⲩ, but not ⲧⲉⲉⲓⲧⲟⲩ) and in the Manichaean MSS ((ⲧ)ⲉⲉⲩ along with (ⲧ)ⲉϊⲧⲟⲩ, but not ⲧⲉ(ⲉ)ⲧⲟⲩ, except Keph 5:33).

(17) The verb ⲁⲙⲁϩⲧⲉ is attested only once, spelled with the same pretonic vowel as in T and in the Manichaean Kephalaia (in contrast to ⲉⲙⲁϩⲧⲉ *A L6 P* and ManiPsalms/Homilies). Other lexemes displaying similar (but not dialectally co-extensive) variation of the pretonic vowel, such as ⲁⲙⲛ̄ⲧⲉ or ⲁϩⲟ, are not attested.

(18) Syllabic gemination of initial ⲛ in ⲛⲁ and ⲛⲏⲩ is not consistent in this MS (as may be expected in a MS displaying such a high degree of inconsistency concerning initial syllabic nasals in general). We find ϩⲛⲛⲁ (2) and ⲛ̄ⲛⲏⲩ (4), but also ⲛⲁ (1) and ⲛⲏⲩ (2). A similar fluctuation is found in T, although the syllabic initial is clearly predominant there (ⲛⲁ and ⲛⲏⲩ only once each) and probably represents the norm of *L5* (cf. the correction made in John 4:25 Thompson), in accordance with dialects *L6 M P*).

(19) The base morph of the plural possessive pronoun is spelled ⲛⲱ= (3) as in T, which is the normal spelling also in *A* and *L4* (in contrast to ⲛⲟⲩ= *L6 F4 S W* ; cf. Funk 1987:52f.).

(20) The imperative of ⲉⲓⲣⲉ (construct form) is attested only once: ⲁⲣⲓ-. In other cases, the imperative marker is either ⲁ-, as in ⲁⲗⲁ= (ⲗⲟ) and ⲁⲗⲁϭⲉ (ⲗⲁϭⲉ) or ⲉ-, as in ⲉϫⲓ= (ϫⲱ); in T ⲉϫⲓ= varies with ⲁϫⲓ=. (No marker is used with ⲛⲉⲩ (11:36) where T has ⲁⲛⲉⲩ). A similar fluctuation is found in MSS of *L6* (where the norm seems to be ⲉ-) and in the Achmimic Clement (ed. Schmidt).

(21) There is ample evidence that the base vowel used in this MS for conjugation bases of Second Present, Conditional and Imperfect is ⲉ- (in accordance with all *L* dialects), without any fluctuation.

(22) The Causative Infinitive is attested only once: prenominal ⲧⲣⲉ- (11:37), in the environment ⲁⲧⲙ-(). This is the form found in *L5*, *L6* and, as a variant, in *L4*.

(23) The form of the Causative Infinitive following the Negative Imperative is not attested in this MS.

(24) The prefix negativing the infinitive is ⲧⲙ- (6), which is once misspelled ⲧⲛ-, but there is no indication that a form ⲧⲙⲛ- (as in *A*) might have been intended.

(25) The interrogative adverb of place is well attested: ⲧⲟ (3), as in T along with *A* and *L4* (in contrast to ⲧⲟⲛ *L6*). Likewise the nasal sonorant seems to be absent in ϣⲙ[ⲧ]- "three' (syllable division in 12:5 excludes a restoration to ϣⲙ[ⲛⲧ]).

(26) The nasal sonorant is always absent from the infix of the Conditional conjugation, which is spelled ϣⲁ-, in accordance with T. (This seems to be a strict normative feature not only of *A*, but also of *L5*; cf. the survey in Funk 1987:50.)

(27) There is no dissimilation of ⲙ in forms such as ⲙⲡⲉ-, ⲙⲡⲁⲧⲉ-. But the nasal in the preposition ϩⲛ- is never assimilated to a following labial consonant, although assimilation is frequent with morphemic ⲛ-. In many cases, however, initial nasals are omitted in this MS (see the preceding chapter). A case of nasal dissimilation (but not initial) is found in ϩⲛⲙⲁⲥ (for ϩⲙⲙⲁⲥ, with nasal gemination).

(28) The Temporal conjugation base is attested predominantly in the usual form with initial nasal sonorant: ⲛ̄ⲧⲁⲣⲉϥ (6), ⲛ̄ⲧⲁⲣⲉⲥ (3), ⲛ̄ⲧⲁⲣⲟⲩ (2), but also without the sonorant: ⲧⲁⲣⲉϥ (2). ⲧⲁⲣⲉⲥ (1). The absence of the sonorant in the present MS, however, is not necessarily due to Achmimic influence, given the general orthographic laxness of the scribe with initial nasals (cf. also the spellings ⲧⲁϥ and ⲧⲁ- for the Relative Perfect, and ⲧⲁⲉⲓ for the Second Perfect).

(29) For a discussion of the form of the proclitic conjunction of closed neutral conditions see above (at the beginning of this chapter). Regarding the presence or absence of initial /i-/, the option of the MS is clearly for absence.

(30) Prenominal base allomorphs are attested only twice for the Circumstantial Present (ⲉ- 11:44; 12:35; perhaps also ⲉⲣⲉ- 11:57, but this is more likely to be a Conditional without ϣⲁ) and four times for the Relative Present (ⲉⲧⲉ- 10:15; 11:2, 32; 12:1; lacuna in 10:40). Thus there is no single safe attestation for the long form with -ⲣⲉ-, which is found in T in a few cases (among these, 10:40), though not in the passages here attested.

(31) The prenominal base allomorph of the Second Present is always ⲉⲣⲉ- 11:13; with ⲛⲁ- (fut.) 11:4, 50; 12:38. To this may be added the prenominal form of the Imperfect ⲛⲉⲣⲉ- 11:5, 8; 12:6, 11; with ⲛⲁ- (fut.) 11:21, 32, 51; as well as the ⲉⲣⲉ- (11:57) mentioned above, as a Conditional. Here the long forms appear to be firmly established.

(32) Undecided fluctuation between the short base allomorph ⲉ⸗ and the long base allomorph ⲉⲣⲉ⸗ when suffixed for the 2nd plural (Circ. and Second Present, Imperfect and Conditional) is a characteristic feature of T. This fluctuation is confirmed by the present MS as a dialectal feature, although this MS does not seem to share the usual practice (found also in T) of combining short base with long suffix or long base with short suffix; instead, it uses the long suffix throughout. Thus we find ⲉⲧⲉⲧⲛ 10:38 (with ⲛⲁ-); 10:38 (with ϣⲁ-) and ⲉⲣⲉⲧⲉⲧⲛ 11:15 (ⲛⲁ- omitted); 12:36 (with ⲛⲁ-). The erroneous spelling ⲉⲣⲉϣⲁ 13:35 (omitting the pronoun) is probably to be emended to ⲉⲣⲉ-ϣⲁ<ⲧⲉⲧⲛ>, although ⲉⲣⲉ<ⲧⲉⲧⲛ>ϣⲁ cannot be excluded. For a survey of these forms see Funk 1987:49f. Note that this MS always spells future forms with two ⲛ: -ⲧⲉⲧⲛⲛⲁ-, whereas T always has ⲛ-simplification: ⲉⲧⲉⲧⲛⲁ and ⲉⲣⲉⲧⲛⲁ.

(33) The use of a long base ϣⲁⲣⲉ⸗ for the Affirmative Aorist (as in *A*) appeared as a feature of *L5* as long as T was its only witness. This feature is, however, not confirmed by the present MS, which has ϣⲁϥ (2) and ϣⲁⲩ (2).

(34/35) The Perfect conjugation base is always ⲁ⸗, never ⲁϩ(ⲁ)⸗.

(36) The 3rd plural form of the possessive articles has always syllabic [u], deleting the unstable ⲉ: ⲡⲟⲩ- (3), ⲛⲟⲩ- (2), in accordance with T. The ⲉ is also deleted in forms such as ⲡⲛ- (2) and ⲧⲥ- (once, alongside with ⲡⲉⲥ- and ⲧⲉⲥ-). The 2nd fem. sing. form is ⲡⲉ-.

(37) The 1st sing. form of the Negative Future is attested only once:

ⲛⲁ 13:8 (lacuna in T, but otherwise also ⲛⲁ in T, as against (ⲛ̄)ⲛⲓ in *L4* and *L6*).

(38) The 1st sing. form of the Causative Infinitive is not attested in this MS.

(39) On account of the deficient representation of the 2nd plural cluster of the Conditional in 13:35 (see above, variable 32) and the lacuna in 13:17, it cannot be said with certainty whether this MS shares the undecided behaviour of T with regard to the linear order of the element ϣⲁ and the suffix; indecision in this case may be a feature of the assumed dialect norm of *L5*. The "normal" order is fully attested only once in a negative form (10:38), but it must be restored also in 13:17 (where T has ⲉⲣⲉϣⲁⲧⲉⲧⲛ) because of the ⲁ preserved after the lacuna.

(40) The allomorph of the relative converter preceding the Affirmative Perfect base is unambiguously attested as ⲛ̄ⲧ- (*L5 L6*) rather than ⲉⲧ- (*L4 A*), although the initial ⲛ has been dropped in a few cases.

(41) The Second Tense allomorph used with the Perfect conjugation is ⲛ̄ⲧ- (as in all *L* dialects and *S*, but in contrast to *A*). As a result of initial n-dropping, in two cases only ⲧ- remains as allomorph.

(42) The prefix of the Negative Imperative is ⲙ̄ⲡⲣ̄- as in all Coptic dialects (notwithstanding spelling variants) except *A* and *P*.

(43) A periphrastic form of the Negative Imperative, combining the absolute (full-stress) form ⲙ̄ⲡⲱⲣ with the nominalized infinitive (with nota infinitivi) does not occur in this MS.

(44) On the peculiar spelling of the proclitic conjunction used in clauses of closed neutral condition see above, at the beginning of this chapter. Regarding the alternative option between -ϫⲉ and -ⲡⲉ, this MS clearly opts for the latter.

(45) The marker of the "Posterior" verb-form in the unconverted conjugation ("First Future") is always ⲛⲁ- (including preformatives represented by an occlusive: ⲕⲛⲁ-, ϥⲛⲁ-, where it is generally ⲁ- in *L4*). With the relative converter, however, the MS has ⲉⲧⲕⲁ- in a single case (11:22).

(46) The "Posterior" marker following the zero-subject Present Relative ⲉⲧ- is ⲛⲁ-, spelled in three cases with double ⲛ to yield a trisyllabic form ⲉⲧⲛⲛⲁ- (i.e., [e.tn.na]). The form ⲉⲧⲁ-, which occurs as variant in so many Subachmimic MSS, is not attested in this MS.

(47) There is no occurrence of a particular "Past participle" formative — whether ⲛ̄ⲧⲁϩ-/ⲉⲧⲁϩ- (as in *L6* and *A*) or ⲉⲣ- (as in T along with *M* and *B4*). Usage of ⲉⲣ-, however, is not overabundant in T,

either, and the few passages where it is used are outside the four chapters covered by the present fragments.

There is at least one variable of greater importance for the distinction of Achmimic/Subachmimic dialects, which should be added to this list, i.e., the form of the demonstrative article of deictic proximity (in opposition to the restricted "weak" demonstrative of non-proximity, ⲡⲓ-/ϯ-/ⲛⲓ-), which is ⲡⲉⲉⲓ-/ⲧⲉⲉⲓ-/ⲛⲉⲉⲓ- in T as well as in the present MS (the same as in Sahidic, with ⲡⲉⲉⲓ- etc. = ⲡⲉϊ- etc.). By contrast, Manichaean Subachmimic (*L4*) joins dialects *A* and *P* in having only one demonstrative article, ⲡⲓ-/ϯ-/ⲛⲓ-, whereas NH-Subachmimic (*L6*) varies between the two series of forms without clear distinction. None of these dialects displays any clear tendency to use the "weak" demonstrative to denote the vague kind of "definiteness" in terms of general knowledge ("higher deixis") or as a kind of "generic" article (as do, in varying degrees, Middle and Lower Egyptian dialects), but *L5* appears to be the only non-Sahidic dialect that follows the well-known Sahidic restrictions for the use of ⲡⲓ-/ϯ-/ⲛⲓ-, while maintaining the three-term distinction of definite determiners: ⲡⲉⲉⲓ- vs. ⲡⲓ- vs. ⲡ(ⲉ)-.

In order to account for the dialectal difference between the present MS and T, at least one other variable has to be added: the subject explication marker, which appears in its Sahidic form ⲛ̄ϭⲓ- in this MS (in accordance with *L6* and the greater part of *A*), whereas it is spelled ϫⲓ- in T (similar to the form ⲛ̄ϫⲓ- *L4*). The attestation of this element is broad enough in both MSS, and there is no variation in either of them (except for the occasional omission of the initial nasal in this MS).

To sum up, there are essentially three features by which the dialect of the present MS is distinguished from T: a greater tendency to vowel raising before /ʔ/ (cf. variables 3-5), the short base of the Aorist conjugation (variable 33), and the Sahidic-type subject explication marker. The second and third of these features render the dialect of this MS more "neutral" (or closer to Sahidic) than that of T at the morphological level, whereas the first feature displays a somewhat closer relationship to Achmimic at the phonological level. This situation is not unsimilar to the kind of in-corpus variation found in other groups of Subachmimic MSS. The number of features that are common to both MSS is great enough for them to form a MS cluster distinct from both *L4* and *L6*. Thus it seems justified, *cum grano salis*, to maintain the hypothesis of a common dialect norm ("*L5*") for the two MSS.

4. The Version

The "version", in the sense of "translation into Coptic", represented by the present MS as well as T is the "Sahidic" version. There are passages where this version is wholly based on a Greek text that deviates from the majority reading, cf., e.g., 10:29 (personal antecedent of relative clause) and 11:33 (the common *varia lectio*); and there are also inner-Coptic peculiarities where all Sahidic MSS go together with the two Subachmimic MSS, cf., e.g., the syntax of 12:46a. In other cases the two Subachmimic MSS share the reading of some of the Sahidic MSS, against others, cf., e.g., 10:15 and 38 ("*my* father"); 11:10 ("no light in *it*", i.e., the night); 11:54 ("to *a* country", indefinite); 12:6b (Imperfect tense). Strikingly, in some cases the two Subachmimic MSS share a particular variant reading preserved in sa 1 (the Palau Ribes codex), which represents the original reading but is not shared by any other Sahidic MS; thus, e.g., 12:42 (**ⲡⲙⲏⲏϣⲉ** without **ⲧⲏⲣϥ**); 11:54 ("among the Jews" instead of "in Judaea", and the maintenance of "no longer"). Agreement between these three MSS can also be found elsewhere, e.g., in 11:48 ("*our* people") and 12:34 (omission through homoioteleuton at the end of the verse; along with Horner's 136).

There are also numerous variant readings, though of no greater importance, peculiar to the two Subachmimic MSS alone (which may be incidental if those Sahidic MSS that had the same reading just happen not to survive), cf., e.g., 10:9 ("through *it*"); 11:2 and 6 (relative clause of Present, not Imperfect); 11:9 (Imperfect: "he saw"); 11:40 (no equivalent for σοι); 11:48 ("*this* place"); 11:57 (**ⲅⲁⲣ**). Some of these peculiarities represent genuine variant readings that are also attested for the Greek, e.g., 11:39 (omission after "Martha"; omission of the vocative "Lord"); 11:54 (no equivalent of ἐκεῖθεν). This "Subachmimic" version is also characterized by several larger omissions in the thirteenth chapter, cf. the text for 13:32 and 13:36; in two other cases, 13:15 and 13:20f. (cf. T), the respective lines are missing in the present MS, but the length of the lacunae between the fragments indicates that this MS probably shared the omissions found in T.

A genuine variant reading peculiar to the present MS is found in Jn 11:18 ("thirty stades" instead of "fifteen stades"). In most other cases where this MS suggests a reading different from all other MSS, this is clearly due to inner-Coptic corruption of the text; thus, e.g., in Jn 12:25 (perhaps the most striking case of corrupt reading in this MS: "he will go up to/beneath it for eternal life" instead of "he will keep it for eternal life").

Finally, I should like briefly to discuss the problem of the "Subachmimic" variant reading ⲉⲧⲃⲉⲟⲩⲉⲉ in 10:33, which apparently stands for περὶ βλασφημίας (which is rendered by some Sahidic MSS as ⲉⲧⲃⲉϫⲓⲟⲩⲁ, thus sa 1 plus Horner's majority, and by others as ϫⲉ ⲉⲕϫⲓⲟⲩⲁ, e.g., sa 4, 5, 9). Based on Thompson's index with reference to this single occurrence in John 10:33, the "Subachmimic" form ⲟⲩⲉⲉ has meanwhile been established as a dialectal variant of ⲟⲩⲁ "blasphemy' throughout the Coptic dictionaries (see Crum, *Dict.* 468b; Westendorf 265; Smith *infra*, Indices). This peculiar reading of T is now confirmed by the MS published here — as is the form ⲟⲩⲁ for the same lexeme in 10:36. Personally, I do not believe that the form ⲟⲩⲉⲉ has any real chance of being a noun meaning "blasphemy" in any of the Subachmimic dialects. The only vocalization to be expected for this lexeme in a non-*FM* dialect is clearly /a/ (phonologically /waʔ/), and the common use of a form with this vocalization in the compound ϫⲉⲓ-ⲟⲩⲁ (ϫⲉ-ⲩⲁ T) in both MSS should be evidence enough that this is indeed the normal spelling of this word in this dialect (cf. also ϫⲉ-ⲟⲩⲁ ManiPs 56:32; 86:16 for *L4*). After all that we know by now about this dialect and its orthography, the form ⲟⲩⲉⲉ in 10:33 can only be the indefinite pronoun, here meaning "something, such a thing", expanded by a ϫⲉ-clause (even though this construction is unusual, instead of the normal ⲡⲉⲉⲓ ... ϫⲉ ...). This "reading" ("but for such a thing ...") may possibly go back to the translation into Subachmimic of a Sahidic source text which had omitted ϫⲓ- before ⲟⲩⲁ, and the resulting Sahidic ⲉⲧⲃⲉⲟⲩⲁ, which still had some chance of meaning "because of blasphemy", was misunderstood by the Subachmimic translator, who substituted his own equivalent of the indefinite pronoun. This would be the first case of evidence from the translational nature of the Subachmimic version to indicate that the direction was indeed from the Sahidic to the Subachmimic (and not vice versa), as it has been established for some Achmimic biblical texts for a long time.

BIBLIOGRAPHY

Černý, J. 1976. *Coptic Etymological Dictionary*. Cambridge: C.U.P.

Crum, W.E. 1934. "Un psaume en dialecte d'Akhmîm", *MIFAO* 67 (*Mélanges Maspero* 2): 73-86.

—. 1937. "The Coptic Glosses", in: F.G. Kenyon, *The Chester Beatty Biblical Papyri ...*, fasc. VI: *Isaiah, Jeremiah, Ecclesiasticus*. London. ix-xii.

Emmel, S. 1981. "Proclitic Forms of the Verb ϯ in Coptic", in: D.W. Young (ed.), *Studies Presented to Hans Jakob Polotsky*, East Gloucester: Pirtle & Polson. 131-146.

Funk, W.-P. 1984. "Bemerkungen zum Sprachvergleich Griechisch-Koptisch", in: P. Nagel (ed.), *Graeco-Coptica.* Martin-Luther-Universität Halle-Wittenberg, Wissenschaftliche Beiträge 1984/48 (I 29). Halle (Saale). 147-180.

—. 1985a. "Koptische Isoglossen im oberägyptischen Raum: 1. ⲉϣϫⲉ 'wenn' etc.," *ZÄS* 112: 19-24.

—. 1985b. "How Closely Related Are the Subachmimic Dialects?", *ZÄS* 112: 124-139.

—. 1986. "Koptische Isoglossen im oberägyptischen Raum: 2. Die Satellitenparadigmen des Präsenssystems", *ZÄS* 113: 103-114.

—. 1987. "Koptische Isoglossen im oberägyptischen Raum: 3. Die Konjugationsformen des Konditional, 4. Die Pluralform des absoluten Possessivpronomens", *ZÄS* 114: 45-54.

Hintze, F., and H.-M. Schenke. 1970. *Die Berliner Handschrift der sahidischen Apostelgeschichte (P. 15 926).* (Texte und Untersuchungen, 109). Berlin: Akademie-Verlag.

Kahle, P.E. 1954. *Bala'izah. Coptic Texts from Deir el-Bala'izah in Upper Egypt.* London: O.U.P.

Kasser, R. 1960. *Papyrus Bodmer VI: Livre des Proverbes.* (CSCO 194/195). Louvain: Peeters.

—. 1981. "Prolégomènes à un essai de classification [systématique] des dialectes et subdialectes coptes selon les critères de la phonétique: III", *Le Muséon* 94: 91-152.

—. 1984. "Orthographe et phonologie de la variété subdialectale lycopolitaine des textes gnostiques coptes de Nag Hammadi", *Le Muséon* 97: 261-312.

—. 1986a. "Causes et procédés de la mutation de ⲛ en aleph ou ⲓ dans trois lexèmes coptes apparentés, *B* ⲱⲛϧ 'vie' et *S* ⲥⲁⲁⲛϣ̄ 'nourrir', *S* ⲡⲣⲁⲛϣ̄ 'archive' (?)", *Orientalia* 55: 177-181.

—. 1986b. "L'identité linguistique du ms. Cambridge Univ. Lib. or. 1700.1 à la périphérie de l'aire lycopolitaine", *Le Muséon* 99: 221-227.

Kuentz, Ch. 1934. "Un cas d'abréviation graphique en copte", *BIFAO* 13: 169-173.

Quecke, H. 1972. *Das Markusevangelium saïdisch.* Barcelona: Papyrologica Castroctaviana.

—. 1984a. *Das Johannesevangelium saïdisch.* Roma-Barcelona: Papyrologica Castroctaviana.

—. 1984b. "Zur Schreibung von i/j in der koptischen Buchschrift", in: *Studien zu Sprache und Religion Ägyptens* (Festschrift W. Westendorf), Göttingen. 289-326.

Rahlfs, A. 1901. *Die Berliner Handschrift des sahidischen Psalters.* (Abhandlungen der Königlichen Gesellschaft der Wissenschaften zu Göttingen. Philologisch-historische Klasse. Neue Folge, Band 4, No. 4). Berlin. (Nachdruck Göttingen: Vandenhoek & Ruprecht, 1970).

Schenke, H.-M. 1985. "Notes on the Edition of the Scheide Codex", in: T. Orlandi and F. Wisse (eds), *Acts of the Second International Congress of Coptic Studies, Roma, 22-26 September 1980.* Rome: C.I.M. 313-322.

SCHMIDT, C. 1925. "Das koptische Didache-Fragment des British Museum", *ZNW* 24: 81-99.

THOMPSON, Sir Herbert. 1924. *The Gospel of St. John According to the Earliest Coptic Manuscript*. London.

VYCICHL, W. 1983. *Dictionnaire étymologique de la langue copte*. Leuven: Peeters.

TRANSCRIPTION WITH NOTES AND CRITICAL APPARATUS

The transcription presents the text as it is found in the MS, adding only word division and the reconstruction of lacunae. Where the scribe's text needs emending, this is dealt with in the apparatus (with the exception of 10:27 and 12:22, where corrections are supplied in the transcription in order to restore complete sentences). In places where the scribe himself made a correction, the transcription presents only what the scribe apparently intended to be there (with one exception: John 10:32). As this is not quite obvious in a number of cases, all correction activities of the scribe are discussed in some detail in the notes below the transcription. Word division is largely reduced to the division of "phonological words", splitting off from such items only those elements (in particular, clitics) which, from a syntactic point of view, do not necessarily share immediate constituency with the rest of the phonological word (e.g., **ⲭⲉ-**, **-ⲡⲉ**).

Apart from emendations and orthographical interpretations, the apparatus mentions any forms differing in at least one letter (not supralinear or other signs) from the MS published by Thompson, except for the following cases of consistent divergence between the two MSS:

- all native Coptic words involving the letter **ϫ**, which are spelled with **ⲭ** throughout this MS, but with **ϫ** in T;
- the consonantal difference in spelling the "subject explication marker" **ⲛ̄ϭⲓ-**, which may be spelled **ⲛ̄ϭⲓ** or **ϭⲓ** in the present MS but is always **ϫⲓ** in T (except 13:3 where it is erroneously spelled **ϫⲉ**);
- the abbreviation **ⲓ̅ⲥ̅**, for which T always has **ⲓ̅ⲏ̅ⲥ̅**;
- the word for "Jew", which is spelled **ⲓⲟⲩⲇⲁⲓ̈ⲉⲓ** in this MS, but **ⲓⲟⲩⲧⲁⲉⲓ** in T.

ABBREVIATIONS

corr. "corrected by the scribe"
del. "to be deleted"
em. "to be emended to"
rest. "restored"
T MS published by Sir Herbert Thompson, *The Gospel of St. John according to the Earliest Coptic Manuscript*, London, 1924
— "lacking in"

For convenience, comments on the lemma and information on T's variant reading have been simplified in the apparatus as far as possible. Thus, for example, an entry "ⲙⲁϥ] i.e. ⲙⲙⲁϥ T" means "the form ⲙⲁϥ is (possibly) meant to be a true spelling variant for ⲙⲙⲁϥ, and T has ⲙⲙⲁϥ"; or an entry "ⲛⲉⲩϩⲛ-] em. ⲛⲉⲩⲛ- T" means "the form ⲛⲉⲩϩⲛ- is to be emended to read ⲛⲉⲩⲛ-, and T has ⲛⲉⲩⲛ-".

p. 4* John 10:7-18

ⲛⲧⲉⲛⲉⲥⲁⲩ ⲟⲩⲁⲛ ⲛⲓⲙ ⲛ̄ⲧⲁⲩⲉⲓ ϩⲉⲛⲥⲁⲛϫⲓⲟⲩⲉ ⲛⲉ 10:7/8
ⲁⲩⲱ ϩⲉⲛⲥⲁⲛⲉ ⲛⲉ ⲁⲗⲗⲁ ⲡⲉⲛ`ⲙⲁⲛ´ⲉⲥⲁⲩ ⲥⲱⲧⲙ ⲁⲣⲁⲩ ⲁ 10:9
ⲛⲁⲕ ⲡⲉ ⲡⲣ̄ⲣⲟ ⲉⲣⲉϣⲁⲟⲩⲉⲉⲓ ⲃⲱⲕ ⲁϩⲟⲩⲛ ⲉ̣ⲓⲧⲟⲟⲧϥ
ϥⲛⲁⲟⲩϫⲉⲉⲓ ⲁⲩⲱ ⲉϥⲛⲁⲃⲱⲕ ⲁϩⲟⲩⲛ ⲛ̄ϥⲓ ⲁⲃⲁⲗ ⲛϥ
ϭⲓⲛⲉ ⲛⲟⲩⲙⲁ̣ⲙ̣ⲁⲛⲉ ⲡⲥⲁⲛϫⲓⲟⲩⲉ ⲙⲁϥⲓ ⲉⲧⲃⲉⲗⲁⲩⲉ 10:10
ⲉⲓⲙⲏⲧⲓ ϫⲉⲕⲁⲥⲉ ⲉϥⲁϫⲓⲟⲩⲉ ⲁⲩⲱ ⲛϥϣⲟⲩⲟⲩⲧ ⲛϥ
ⲥϯ̣ⲥ̣ⲉⲕⲟ ⲁⲛⲁⲕ ⲛⲧⲁⲉⲓ ϫⲉⲕⲁⲥⲉ ⲉϥⲛⲁϫⲓ ⲛⲟⲩⲱⲱϩ ⲁⲩⲱ
ⲛ̄ⲥⲉϫⲓⲟⲩϩⲟ ⲁⲛⲁⲕ ⲡⲉ ⲡϣ̄ⲥ̄ ⲉⲧⲛⲁⲛⲟⲩϥ ⲡϣⲥ ⲉⲧⲛⲁ 10:11
ⲛⲟⲩϥ ϣⲁϥⲕⲟⲩ ⲧⲉⲯ[ⲩ]ⲭⲏ ϩⲁⲛⲛⲉϥⲉⲥⲁⲩ ⲡϫⲁⲉⲓⲃⲉⲕⲉ 10:12
ⲇⲉ ⲉⲧⲉ ⲟⲩϣⲱⲥ ⲉⲛ̣ ⲡ̣ⲉ ⲡⲉⲉⲓ ⲉⲧⲉ ⲛⲱϥ ⲉⲛ ⲛⲉ ⲛⲉⲥⲁⲩ
ϣⲁϥⲛⲉⲩ ⲁⲡⲟⲩⲱⲛϣ ⲉϥⲛⲏⲩ ⲁⲩⲱ ⲛϥⲕⲁⲛⲉⲥⲁⲩ
ⲛ̄ϥⲡⲱⲧ ⲛⲧⲉⲡⲟⲩⲱ̣ϣⲛϣ ⲧⲁⲣⲡⲟⲩ ⲁⲩⲱ ⲛϥϫⲁⲣⲟⲩ
ⲁⲃⲁⲗ ϫⲉ ⲟⲩϫⲁⲉⲓⲃⲉⲕⲉ ⲡⲉ ⲁⲩⲱ ⲡⲉϥⲣⲁⲟⲩϣ ⲉⲛ ⲡⲉ 10:13
ϩⲁⲛⲛⲉⲥⲁⲩ ⲁⲛⲁⲕ ⲡⲉ ⲡϣⲥ ⲉⲧⲛⲁⲛⲟⲩϥ ⲁⲩⲱ ϯⲥⲁⲩ 10:14
ⲛⲉⲉ ⲛⲛⲱⲉⲓ ⲁⲩⲱ ⲛⲱⲉⲓ ⲥⲉⲥⲁⲩⲛⲉ ⲙⲁⲉⲓ ⲕⲁⲧⲁⲑⲉ ⲉⲧⲉ 10:15
ⲡⲁⲉⲓⲱⲧ ⲥⲁⲩⲛⲉ ⲙⲁⲉⲓ ⲁⲩⲱ ⲁⲛⲁⲕ ϯⲥⲁⲩⲛⲉ ⲡⲁⲉⲓⲱⲧ

10:8 11 The scribe originally wrote ⲛⲉⲥⲁⲩ but then added ⲙⲁⲛ above the line, thus changing "the sheep" into "the shepherds" (which makes poor sense here).

10:10 16 In what has been transcribed as ⲥϯ̣ⲥ̣ⲉⲕⲟ there is some blotting between ⲥ and ⲉⲕⲟ, which makes it difficult to say what has been intended to be there by the scribe (probably not ⲥϯⲥⲉⲕⲟ). Although the vestiges of letters now discernible look like ϯⲥ, the blotted ink may be due to a correction by the scribe, who either crossed out the two letters (to read ⲥⲉⲕⲟ) or imposed the shape of a thick ⲧ on the place occupied by the two letters (to read ⲥⲧⲉⲕⲟ); for the latter form cf. ⲥⲧⲉⲃⲱⲧⲛ in 10:32.

---

10 ⲛⲧⲉ-] ⲛ- T 11 ⲡⲉ-] i.e. ⲙⲡⲉ- T | ⲛⲙⲁⲛⲉⲥⲁⲩ corr.] ⲛⲉⲥⲁⲩ T 12 ⲡⲣⲣⲟ] i.e. ⲡⲣⲟ T | ⲟⲩⲉⲉⲓ] em. ⲟⲩⲉⲉ T | ⲉⲓⲧⲟⲟⲧϥ] i.e. ϩⲓⲧⲟⲟⲧϥ T (corr. ϩⲓⲧⲟⲟⲧ?) 13 ⲉϥⲛⲁ-] ϥⲛⲁ- T 14 ⲙⲁⲙⲁⲛⲉ] i.e. ⲙⲁ ⲙⲙⲁⲛⲉ T | ⲙⲁϥⲓ] ⲙⲁϥⲉⲓ T 15 ϣⲟⲩⲟⲩⲧ] ϣⲱⲱⲧ+ⲁⲩⲱ T 16 ⲥϯⲥⲉⲕⲟ (uncertain; see note above)] ⲧⲥⲉⲕⲟ T | ⲛⲧⲁⲉⲓ] T; i.e. ⲛⲧⲁⲉⲓⲉⲓ | ⲉϥⲛⲁ-] em. ⲉⲩⲛⲁ-; ⲉⲩⲁ- T | ⲱⲱϩ] i.e. ⲱⲱⲛϩ T 17 ⲟⲩϩⲟ] em. ⲟⲩϩⲟⲩⲟ T | ϣⲥ] bis, i.e. ϣⲱⲥ T 18 ϣⲁϥ-] ϣⲁⲣⲉϥ- T | ⲕⲟⲩ ⲧⲉ-] em. ⲕⲟⲩ ⲛⲧⲉ- or ⲛⲧⲉϥ-; ⲕⲁⲧⲉϥ- T | ϩⲁⲛ-] em. ϩⲁ- T 20 ϣⲁϥ-] ϣⲁⲣⲉϥ- T | ⲟⲩⲱⲛϣ] ⲟⲩⲱⲱⲛϣ corr. T | ⲛⲏⲩ] ⲛⲛⲏⲩ T 21 ⲛⲧⲉ-] ⲁⲩⲱ ϣⲁⲣⲉ- T | ⲟⲩⲱϣⲛϣ] em. ⲟⲩⲱⲱⲛϣ T, or ⲟⲩⲱⲛϣ | ⲁⲩⲱ] — T 22 ⲡⲉϥ-] ⲡϥ- rest. T 23 ϩⲁⲛⲛ-] i.e. ϩⲁⲛ- T | ϣⲥ] i.e. ϣⲱⲥ T 23/24 ϯⲥⲁⲩⲛⲉⲉ] em. ϯⲥⲁⲩⲛⲉ T 24 ⲥⲉ-] — T | ⲙⲁⲉⲓ] i.e. ⲙⲙⲁⲉⲓ T 25 ⲙⲁⲉⲓ] i.e. ⲙⲙⲁⲉⲓ T | ⲡⲁⲉⲓⲱⲧ[2]] em. ⲙⲡⲁⲉⲓⲱⲧ T

ⲁⲩⲱ ϯⲛⲁⲕⲟⲩ ⲛⲧⲁⲯⲩⲭⲏ ϩⲁⲛⲉⲥⲁⲩ ⲟⲩⲛⲧⲏⲓ ⲇⲉ 10:16
ⲁⲛ ⲙ̄ⲙⲉⲩ ϩⲉⲛⲕⲉⲉⲥⲁⲩ ⲉϩⲉⲛⲁⲃⲁⲗ ⲉⲛⲉ ⲙ̄ⲡⲓ
ⲁϩⲉⲥⲁⲩ ⲁⲩⲱ ⲛⲉⲧⲛⲙⲉⲩ ⲁⲛ ϯⲛⲁⲥⲁ`ⲩ´ϩⲟⲩ ⲛ̄ⲥⲉ
ⲥⲱⲧⲙ ⲁⲧⲁⲥⲙⲏ ⲛ̄ⲥⲉϣⲱⲡⲉ ⲛ̄ⲟⲩⲉⲓⲁϩⲉⲥⲁⲩ ⲛ̄
ⲛⲟⲩⲱⲧ ϣⲱⲥ ⲛ̄ⲟⲩⲧ ⲉⲧⲃⲉⲡⲉⲉⲓ ⲡⲁⲉⲓⲱⲧ ⲙⲁⲉⲓ 10:17
ⲉ ⲙⲁⲉⲓ ϫⲉ ϯⲛⲁⲕⲟⲩ ⲛⲧⲁⲯⲩⲭⲏ ϫⲉⲕⲁⲥⲉ ⲁⲛ
ⲉⲓⲛⲁϫⲓⲧⲥ ⲙⲛ̄ⲗⲁ̣ⲩⲉⲉ ϥⲓ ⲙⲁⲥ ⲛ̄ⲧⲟⲟⲧ ⲁⲗⲗⲁ 10:18

10:17 31 The letters ϯ and ⲯ in ϯⲛⲁⲕⲟⲩ and ⲛⲧⲁⲯⲩⲭⲏ, respectively, are indistinguishable in shape.

---

26 ϩⲁⲛ-] ϩⲁⲛⲁ- T | ⲟⲩⲛⲧⲏⲓ] ⲟⲩⲛⲧⲏⲉⲓ T 27 ϩⲉⲛ-] em. ⲛϩⲉⲛ- T | ⲉⲛⲉ] em. ⲉⲛ ⲛⲉ T 27/28 ⲙⲡⲓⲁϩ-] ⲙⲡⲉⲉⲓⲁϩ- (i.e. ⲙⲡⲉⲉⲓ-ⲉⲓⲁϩ-, *pace* Thompson) T 28 ⲛⲉⲧⲛⲙⲉⲩ] i.e. ⲛⲉⲧⲙⲙⲉⲩ T | ⲥⲁⲩϩⲟⲩ corr.] ⲥⲁⲟⲩϩⲟⲩ T 29/30 ⲛⲛ-] i.e. ⲛ- T 30 ϣⲱⲥ] em. ⲟⲩϣⲱⲥ T | ⲟⲩⲧ] i.e. ⲟⲩⲱⲧ T 31 ⲙⲁⲉⲓ] i.e. ⲙⲙⲁⲉⲓ T 32 ⲉⲓⲛⲁ-] ⲉⲉⲓⲛⲁ- T | ⲗⲁⲩⲉⲉ] em. ⲗⲁⲩⲉ T | ⲙⲁⲥ] i.e. ⲙⲙⲁⲥ T

p. 5* John 10:18-36

ⲁⲛⲁⲕ ⲉⲧⲕⲱ ⲙⲁⲥ ϩⲁⲣⲁⲉⲓ ⲟⲩⲉⲧ [ⲟⲩⲛϯⲧⲉⲝⲟⲩⲥⲓ]ⲁ̣
ⲁⲕⲁⲁⲥ ⲁⲩⲱ ⲟⲩⲛϯⲧⲉⲝⲟⲩⲥⲓⲁ ⲁϫ̣ⲓ̣ⲧ̣[ⲥ ϯⲉⲛⲧ]ⲟ̣ⲗ̣ⲏ
ⲧⲁⲉⲓϫⲓⲧⲥ ⲁⲃⲁⲗ ϩⲓⲧⲛⲡⲁⲉⲓⲱⲧ ⲁⲩⲱ ⲁ[ⲟⲩⲥⲭⲓⲥ]ⲙⲁ 10:19
ϣⲱⲡⲉ ϩⲛⲛⲓⲟⲩⲇⲁⲉⲓ ⲉⲧⲃⲉⲛⲛⲉⲉⲓⲥⲉϫⲉ [ⲛⲉⲩⲛϩ]ⲁ̣ϩ̣ ⲇⲉ 10:20
ϫⲱ ⲙⲙⲁⲥ ϩⲏⲧⲟⲩ ϫⲉ ⲟⲩⲇⲁⲓⲙⲟⲛⲓⲟⲛ ⲡ[ⲉⲧⲛⲙⲙⲉ]ϥ̣
ⲁⲩⲱ ⲉϥⲗⲁⲃⲉ ⲉⲧⲃⲉⲉⲩ ⲧⲉⲧⲛⲥⲱⲧⲙ ⲁⲣⲁϥ ϩⲉ[ⲛⲕⲁⲩⲉ] ⲇ̣ⲉ̣ ⲉⲩ 10:21
ϫⲱ ⲙⲙⲁⲥ ϫⲉ ⲛⲉⲉⲓⲥⲉϫⲉ ⲛⲁⲟⲩⲉ ⲉⲛ̣ ⲛ̣[ⲉ ⲉϥⲟ ⲛ]ⲇⲁⲓ
ⲙⲱⲛ ⲙⲏ ⲟⲩⲛϣϭⲁⲙ ⲛ̄ⲇⲁⲓⲙⲟⲛⲓⲟⲛ̣ [ⲁⲧⲣⲉϥⲟⲩ]ⲉ̣ⲛ ⲁⲛ
ⲃⲉⲗ ⲛ̄ⲟⲩⲃⲗⲗⲉ ⲁϥϣⲱⲡⲉ ⲇⲉ ϩⲛⲡⲟⲩⲁ[ⲉⲓϣ ⲉⲧⲛⲙⲉ]ⲩ ⲛ 10:22
ϭⲓⲡϫⲓⲁⲉⲓⲕ ϩⲛⲑⲏⲉⲣⲟⲩⲥⲁⲗⲏⲙ ⲛⲉⲧⲡⲣ[ⲱ ⲧⲉ ⲁⲩ]ⲱ̣ ⲛⲉϥ 10:23
ⲙⲁϩⲉ ⲛ̄ϭⲓⲓ̅ⲥ̅ ϩⲛⲡⲣ̄ⲡⲉⲉ ϩⲁⲥⲧⲟⲁ ⲥⲟⲗ[ⲟⲙⲱⲛ] ⲁ̣ⲩ`ⲕ´ⲱ 10:24
ⲧⲉ ϭⲉ ⲁⲣⲁϥ ϭⲓⲛⲓⲟⲩⲇⲁⲉⲓ ⲡⲁϫⲉⲩ ⲛⲉϥ ϫⲉ ϣ̣[ⲁⲛⲧⲉⲉⲩ ϣⲱⲡⲉ]
ⲉⲕϥⲓ ⲡⲉⲕϩⲏⲧ ⲉϣⲱⲡⲉ ⲧⲁⲕ ⲡⲉ ⲡⲉⲭ̣̅ⲥ̅ ⲉϫⲓⲥ ⲛⲉⲛ [ϩ]ⲛⲟⲩ
ⲡⲁⲣⲏⲥⲓⲁ ⲁϥⲟⲩⲱϣⲃ ⲛⲉⲩ ⲛ̄ϭⲓⲓ̅ⲥ̅ ϫⲉ ⲁⲉⲓϫⲟⲟⲥ ⲛⲏⲧⲛ 10:25
ⲁⲩⲱ ⲧⲉⲧⲛⲣⲡⲓⲥⲧⲉⲩⲉ ⲉⲛ ⲁⲣⲁⲉⲓ ⲛ̄ϩⲃⲏⲩⲉ ⲁⲛⲁⲕ ϯⲉⲓⲣⲉ
ⲙⲙⲁⲩ ϩⲛⲡⲣⲉⲛ ⲡⲁⲉⲓⲱⲧ ⲛⲉⲉⲓ ⲛⲉ ⲉⲧⲣⲙⲛⲧⲣⲉ ⲉⲧⲃⲏⲧ
ⲁⲗⲗⲁⲁ ⲛⲧⲱⲧⲛ ⲧⲉⲧⲛⲣⲡⲓⲥⲧⲉⲩⲉ ⲉⲛ ϫⲉ ⲛⲧⲱⲧⲛϩⲉⲛⲁⲃⲁⲗ 10:26
ⲉⲛ ϩⲛⲛⲁⲉⲥⲁⲩ <ⲛⲁⲉⲥⲁⲩ> ⲁⲛⲁⲕ ϣⲁⲩⲥⲱⲧⲙ ⲁⲧⲁⲥⲙⲏ ⲁⲩⲱ 10:27
ϯⲥⲁⲩ
ⲛⲉ ⲙⲙⲁⲩ ⲛ̄ⲧⲁⲩ ⲇⲉ ⲥⲉⲛⲁⲟⲩⲁϩⲟⲩ̣ ⲥⲱⲉⲓ ⲁⲩⲱ ⲁⲛⲁⲕ 10:28

10:25 14 In ⲛ̄ϭⲓⲓ̅ⲥ̅ there is a wide empty space between ⲛ̄ and ϭⲓⲓ̅ⲥ̅.

10:26/27 18 As regards the haplography of ⲛⲁⲉⲥⲁⲩ, the shape of the last letter ⲩ (with the upstroke extending far to the right), otherwise mainly found at the end of a line, suggests that the scribe meant to finish the sentence here and thus to start the new sentence with ⲁⲛⲁⲕ.

---

1 ⲙⲁⲥ] i.e. ⲙⲙⲁⲥ T | ⲟⲩⲉⲧ] i.e. ⲟⲩⲁⲉⲧ; ⲟⲩⲁⲉⲉⲧ T (partly rest.) 3 ⲧⲁⲉⲓ] i.e. ⲛⲧⲁⲉⲓ- T | ⲥϫⲓⲥⲙⲁ rest.] ⲥϫⲓⲥⲙⲁ T 4 ⲉⲧⲃⲉⲛⲛⲉⲉⲓ-] i.e. ⲉⲧⲃⲉⲛⲉⲉⲓ- T 5 ϩⲏⲧⲟⲩ] i.e. ⲛϩⲏⲧⲟⲩ T 6 ⲇⲉ] — T 7 ⲟⲩⲉ] ⲟⲩⲉⲉ T 8 ⲟⲩⲛϣ-] ⲟⲩⲛ- T | ⲛⲇⲁⲓⲙⲟⲛⲓⲟⲛ] ⲛⲟⲩⲇⲁⲓⲙⲱⲛ T | ⲁⲧⲣⲉϥ- rest.] ⲁ- T 9 ⲉⲧⲛⲙⲉⲩ rest.] ⲉⲧⲙⲙⲉⲩ T 10 ⲑⲏⲉⲣⲟⲩⲥⲁⲗⲏⲙ] i.e. ⲑⲓⲉⲣⲟⲩⲥⲁⲗⲏⲙ T 11 ⲙⲁϩⲉ] ⲙⲁⲁϩⲉ+ⲡⲉ T | ϩⲁⲥⲧⲟⲁ] ϩⲁⲧⲥⲧⲟⲁ ⲛ- T 12 ϭⲓ-] i.e. ⲛϭⲓ- 13 ⲡⲉⲕϩⲏⲧ] em. ⲙⲡⲛϩⲏⲧ T | ⲉϣⲱⲡⲉ] ⲉϣⲡⲉ T | ⲧⲁⲕ] i.e. ⲛⲧⲁⲕ T | ⲡⲉⲭ̅ⲥ̅] ⲡⲭ̅ⲥ̅ T 14 ⲡⲁⲣⲏⲥⲓⲁ] ⲡⲁⲣⲣⲏⲥⲓⲁ T 15 ϯ-] em. ⲉϯ- T 16 ⲡⲁ-] em. ⲙⲡⲁ- T | ⲛⲉ ⲉⲧ-] ⲛⲉⲧ- T 17 ⲁⲗⲗⲁⲁ] em. ⲁⲗⲗⲁ T 18 ϩⲛⲛⲁⲉⲥⲁⲩ MS] ϩⲛⲛⲁⲉⲥⲁⲩ· ⲛⲁⲉⲥⲁⲩ T (partly rest.) | ϣⲁⲩ-] ϣⲁⲣⲟⲩ- rest. T 19 ⲥⲱⲉⲓ] i.e. ⲛⲥⲱⲉⲓ T

ϯⲛⲁϯ ⲛⲉⲩ ⲛⲟⲩⲱⲱϩ ⲛ̄ϣⲁⲁ̣ⲛⲏϩⲉ ⲛ̄ⲥⲉⲧⲙϩⲁⲉⲓⲉ ⲁⲃⲁⲗ
ⲁⲛⲏϩⲉ ⲁⲩⲱ ⲙⲧⲉ`ⲧⲙ´ ⲗⲁⲩⲉ ⲧⲁⲣⲡⲟⲩ ⲁⲃⲁⲗ ϩⲛⲛⲁϭⲓϫ ⲡⲁ 10:29
ⲉⲓⲱⲧ ⲛ̄ⲧⲁϥⲧⲉⲉⲓⲧⲟⲩ ⲛⲏⲉⲓ ⲉϥⲓϫⲛⲟⲩⲁⲛ ⲛⲓⲙ ⲁⲩⲱ ⲙ̄ⲛϭⲁⲙ̣
ⲛ̄ⲗⲁⲩⲉ ⲁⲧⲁⲣⲡⲟⲩ ⲁⲃⲁⲗ ϩⲛⲛϭⲓϫ ⲙⲡⲁⲉⲓⲱⲧ̇ ⲁⲛⲁⲛ 10:30
ⲟⲩⲉⲉ ⲁⲩϥⲓⲱⲛⲉ ⲁⲛ ⲛ̄ϭⲓⲛⲓⲟⲩⲇⲁⲉⲓ ϫⲉⲕⲁⲥⲉ ⲉⲩⲁⲛⲟ[ⲩ] 10:31
ϫⲉ ⲁⲣⲁϥ ⲁϥⲟⲩⲱϣⲃ ⲛⲉⲩ ⲛ̄ϭⲓⲓ̅ⲥ̅ ϫⲉ ⲁⲉⲓⲥⲧⲉⲃⲱⲧⲛ ⲁϩ[ⲁϩ] 10:32
ⲛϩⲱⲃ ⲉⲛⲁⲛⲟⲩϥ ⲁⲃⲁⲗ ϩⲓⲧⲛⲡⲁⲉⲓⲱⲧ̇ ⲉⲧⲃⲉⲉϣ ⲙⲙ[ⲁⲩ]
ⲛ̄ϩⲱⲃ {ⲉⲛⲁⲛⲟⲩϥ ⲁⲃⲁⲗ ϩⲓⲧⲛⲡⲁⲉⲓⲱⲧ} ⲉⲧⲉⲧⲛⲛ[ⲁ]
ϫⲱⲛⲉ ⲁⲣⲁⲉⲓ ⲁⲩⲟⲩⲱϣⲃ ⲛⲉϥ ⲛ̄ϭⲓⲛⲓⲟⲩⲇⲁⲉⲓ ϫⲉ ⲉ̣[ⲛⲁ] 10:33
ⲛⲁϫⲱⲛⲉ ⲉⲛ ⲁⲣⲁⲕ ⲉⲧⲃⲉⲱϣ ⲉⲛⲁⲛⲟⲩϥ ⲁⲗⲗⲁ ⲉⲧ̣[ⲃⲉ]
ⲟⲩⲉⲉ ϫⲉ ⲛⲧⲕⲟⲩⲣⲱⲙⲉ ⲉⲕⲓⲣⲉ ⲙⲁⲕ ⲛ̄ⲛⲟⲩⲧⲉ
ⲁϥⲟⲩⲱϣⲃ ⲛⲉⲩ ⲛ̄ϭⲓⲓ̅ⲥ̅ ϫⲉ ⲙⲏ ϥⲥⲏϩ ⲉⲛ ϩⲛⲡⲉⲧⲛ̣ 10:34
ⲛⲟⲙⲟⲥ ϫⲉ ⲁⲛⲁⲕ ⲁⲉⲓϫⲟⲟⲥ ϫⲉ ⲛⲧⲱⲧⲛϩⲉⲛⲛ[ⲟⲩ]
ⲧⲉ ⲉϣⲱⲡⲉ ⲁϥⲙⲟⲩⲧⲉ ⲁⲛⲉⲧⲛⲙⲉⲩ ϫⲉ ⲛⲟ[ⲩⲧⲉ] 10:35
ⲛⲉⲉⲓ ⲛⲧⲁⲛⲥⲉϫⲉ ⲙⲡⲛⲟⲩⲧⲉⲉ ϣⲱⲡⲉ ϣⲁ[ⲣⲁⲩ]
ⲁⲩⲱ ⲙⲛϭⲁⲙ ⲛⲧⲉⲧⲅⲣⲁⲫⲏ ⲃⲱⲗ ⲁⲃⲁⲗ ⲡ̣[ⲉⲉⲓ] 10:36

10:28 21 ⲧⲙ is written above the ⲧ of ⲙⲧⲉ; the scribe did not bother to change ⲙⲧⲉ into ⲛⲧⲉ.

10:30 23 In ⲁⲛⲁⲛ the second ⲛ is written in bold strokes over previous ⲕ, the oblique strokes of which are still visible. Thus the scribe first wrote correctly ⲁⲛⲁⲕ but failed to continue with ⲙⲛⲡⲁⲉⲓⲱⲧ, probably because he realized that he had just written ⲡⲁⲉⲓⲱⲧ. He mistook this for the second ⲡⲁⲉⲓⲱⲧ, and his former ⲁⲛⲁⲕ for an error.

10:32 25 Instead of ⲛⲉⲩ the scribe first had written ⲛⲉϥ, but then imposed a short oblique stroke over the last letter, thus changing ϥ into something like ⲩ (not the usual shape). — In ⲛ̄ϭⲓⲓ̅ⲥ̅, the second stroke extends above the letters ⲓⲥ.

10:32 26/27 In the MS, the words ⲉⲧⲃⲉⲉϣ ⲙ̄ⲙⲁⲩ ⲛ̄ϩⲱⲃ ⲉⲛⲁⲛⲟⲩϥ ⲁⲃⲁⲗ ϩⲓⲧⲛ-ⲡⲁⲉⲓⲱⲧ are cancelled by a surrounding broken line. Obviously, after writing this ⲛ̄ϩⲱⲃ, the scribe had slipped back in his exemplar to the ⲛ̄ϩⲱⲃ of the preceding sentence and copied all the rest up to ϩⲓⲧⲛⲡⲁⲉⲓⲱⲧ for a second time. When he realized his mistake, he cancelled all the words from as far back as the first ϩⲓⲧⲛⲡⲁⲉⲓⲱⲧ, thus wrongly including ⲉⲧⲃⲉⲉϣ ⲙ̄ⲙⲁⲩ ⲛ̄ϩⲱⲃ in the cancellation.

Bottom margin of p. 5*: The letters ⲣⲟ̣ⲍ (ⲟ could also be ⲑ) are written a little right of the centre of the bottom margin. Although not distinguished by any stroke, this was probably meant to be a number. Its actual significance is unclear: it cannot be a quire or page number, and as a stichometric figure, 177 does not seem to make sense just here.

---

20 ⲱⲱϩ] i.e. ⲱⲱⲛϩ rest. T | ⲛϣⲁ-] prob. ϣⲁ- T | ϩⲁⲉⲓⲉ] apparently ϩⲉⲉⲓⲉ to be restored in T 21 ⲁⲛⲏϩⲉ] ϣⲁⲁⲛⲏϩⲉ T | ⲙⲧⲉⲧⲙ- corr.] em. ⲛⲧⲉⲧⲙ-; ⲛⲉ- rest. T 22 ⲧⲉⲉⲓⲧⲟⲩ] ⲧⲉⲉⲩ T (partly rest.) | ⲉϥⲓϫⲛ-] i.e. ⲉϥϩⲓϫⲛ-; ϩⲓϫⲛ- T 23 ⲁⲛⲁⲛ-] em. ⲁⲛⲁⲕ ⲙⲛⲡⲁⲉⲓⲱⲧ ⲁⲛⲁⲛ- T 24/25 ⲛⲟⲩϫⲉ] ⲛⲁϫⲟⲩ T 25 ⲥⲧⲉ-ⲃⲱⲧⲛ] ⲧⲥⲉⲃⲱⲧⲛ T 26 ⲉⲧⲃⲉⲉϣ etc. (corr.)] see note above 27 ⲉⲛⲁⲛⲟⲩϥ ⲁⲃⲁⲗ ϩⲓⲧⲛⲡⲁⲉⲓⲱⲧ] — T 29 ⲱϣ] ϩⲱⲃ T 30 ⲉⲕ-] ⲕ- T | ⲙⲁⲕ] i.e. ⲙⲙⲁⲕ T 33 ⲉϣⲱⲡⲉ] ⲉϣⲡⲉ T (partly rest.) | ⲛⲉⲧⲛⲙⲉⲩ] i.e. ⲛⲉⲧⲙⲙⲉⲩ rest. T 34 ⲛⲥⲉϫⲉ] ⲡⲥⲉϫⲉ rest. T | ⲛⲟⲩⲧⲉⲉ] em. ⲛⲟⲩⲧⲉ T | ϣⲁⲣⲁⲩ rest.] ϣⲁⲁⲣⲁⲩ rest. T 35 ⲁⲃⲁⲗ+ϩⲓⲉ T 35/1 ⲡⲉⲉⲓ (rest.) ⲛⲧⲁ-] ⲡⲉⲛⲧⲁ- T

p. 6* John 10:36-11:12

ⲛ̅ⲧⲁⲡⲓⲱⲧ̇ [ⲧⲟⲩⲃⲁϥ] ⲁϥⲧⲛⲛⲁⲟⲩϥ ⲁⲡⲕⲟⲥⲙⲟⲥ ⲛ̅ⲧⲱⲧⲛ
[ⲧ]ⲉⲧⲛϫⲱ ⲙ[ⲁⲥ ⲁⲣⲁϥ] ϫⲉ ⲉⲕϫⲉⲓⲟⲩⲁ ϫⲉ ⲁⲉⲓϫⲟⲟⲥ ϫⲉ ⲁⲛⲁⲕ
ⲡ̣ⲉ ⲡϣ[ⲏⲣⲉ ⲙⲡⲛⲟ]ⲩ̣ⲧⲉ ⲉϣⲱⲡⲉ ϯⲉⲓⲣⲉ ⲛ̅ⲃⲏⲩⲉ ⲉⲛ ⲡⲁⲉⲓ 10:37
ⲱ̣ⲧ ⲙ̅ⲡⲣⲣ̣[ⲡⲓⲥⲧⲉⲩ]ⲉ̣ ⲁⲣⲁⲉⲓ ⲉϣⲱⲡⲉ ϯⲉⲓⲣⲉ ⲙⲁⲩ ⲕⲁⲛ ⲉⲧⲉ 10:38
ⲧ̣ⲛϣⲁⲧⲙⲣ̣[ⲡⲓⲥⲧⲉⲩ]ⲉ ⲁⲣⲁⲉⲓ ⲁⲣⲓⲡⲓⲥⲧⲉⲩⲉ ϩⲁⲛⲃⲏⲩⲉ ϫⲉ
ⲕ̣ⲁⲥⲉ ⲉⲧⲉⲧⲛⲛ̣[ⲁⲙⲙⲉ ⲁ]ⲩ̣ⲱ ⲧⲉⲧⲛⲥⲁⲩⲛⲉ ϫ`ⲉ´ ⲡⲁⲉⲓⲱⲧ ϣⲟⲡ ϩⲏⲧ
[ⲁ]ⲩⲱ ⲁⲛⲁ̣[ⲕ ϩⲣⲏⲓ ϩ]ⲏ̣ⲧϥ ⲛⲉⲩϣⲓ ⲇⲉ ⲛ̅ⲥⲁⲙⲁϩⲧⲉ ⲙⲁϥ ⲁⲩⲱ 10:39
[ⲁϥⲓ ⲁ]ⲃ̣ⲁⲗ ϩ[ⲛⲧⲟⲩϭⲓ]ϫ ⲁϥⲃⲱⲕ ⲁⲛ ⲁⲡⲓⲕⲣⲟ ⲡⲓⲟⲣⲇⲁⲛⲏ̣ⲥ̣ 10:40
[ⲁⲡⲙ]ⲁ̣ ⲉⲧ[ⲉⲣⲉⲓⲱⲁⲛ]ⲛⲏⲥ ϩⲏⲧϥ ⲛϣⲁⲣⲉⲡ ⲉϥⲣ̅ⲃⲁⲡⲧⲓⲍⲉ
[ⲁⲩⲱ ⲁϥϭⲟⲩ ⲙⲙⲉⲩ ⲁⲩ]ⲙⲏϣⲉ ⲃⲱⲕ ϣⲁⲁⲣⲁϥ ⲁⲩⲱ ⲛⲉⲩϫⲱ̣ 10:41
ⲙ̣ⲙⲁ̣[ⲥ ϫ]ⲉ̣ [ⲙⲡⲉⲓⲱ]ⲁ̣ⲛⲛⲏⲥ ⲙⲉⲛ ⲣ̅ⲗⲁⲩⲉ ⲙⲙⲁⲉⲓⲛ ⲥⲉϫⲉ
[ⲛ]ⲓⲙ ⲛ̅ⲧⲁⲉⲓⲓⲱⲁⲛⲛⲏⲥ ϫⲟⲟⲩ ⲉⲧⲃⲉⲡⲉⲉⲓ ϩⲉⲛⲙⲓⲉ̣ ⲛⲉ ⲁⲩⲱ 10:42
ⲁ̣ϩⲁϩ ⲣⲡⲓⲥⲧⲉⲩⲉ ⲁⲣⲁϥ ⲛⲉⲩϩⲛⲟⲩⲉⲉⲉ ⲇⲉ ϣⲱⲛⲉ ϫⲉ ⲗⲁ 11:1
ⲍⲁⲣⲟⲥ ⲁⲃⲁⲗ ϩⲛⲃⲏⲑⲁⲛⲓⲁⲁ ⲡϯⲙⲉ ⲙⲁⲣⲓⲁⲙ ⲙⲛⲙⲁⲣ
ⲑⲁ ⲧⲉⲥ̣ⲥⲱⲛⲉⲉ ⲧⲉ̣ⲉ̣ⲓ̣ ⲧ̣ⲉ ⲙⲁⲣⲓⲁ ⲛ̅ⲧⲁⲥⲧⲁϩⲁⲥⲡϫⲁⲉⲓⲥ 11:2
ⲛ̣̅ⲥⲁϭⲛ ⲁⲩⲱ ⲁⲥϥⲛⲧⲛⲉϥⲟⲩⲣⲓⲧⲉ ⲙⲡⲉⲥϥⲟⲩⲉ ⲧⲉ
[ⲉ]ⲓ ⲧⲉ ⲡⲉⲥⲁⲛ ⲗⲁⲍⲁⲣⲟⲥ ϣⲱⲛⲉ ⲁⲩϫⲁⲩ ϭⲉ ϣⲁⲣⲁϥ ⲛ̅ 11:3
[ϭⲓ]ⲛⲉϥⲥⲱⲛⲉ ⲉⲩϫⲱ ⲙⲙⲁⲥ ϫⲉ ⲉⲓⲥϩⲏⲡⲉ ⲉϥϣⲱⲛⲉ
[ⲛ]ϭⲓⲡⲉⲧⲉⲕⲙⲁⲉⲓⲉ̣ ⲙⲁϥ ⲛ̅ⲧⲁⲣⲉϥⲥⲱⲧⲙ ϭⲉ ϭⲓⲓ̅ⲥ̅ 11:4

10:38 6 The scribe wrote ⲥⲁⲩⲛⲉϫ ⲡⲁⲉⲓⲱⲧ, with a large initial ⲡ whose horizontal bar extends into the small-size ϫ. Above this extension, he then added ⲉ.

10:38 7 In the lacuna, there is space for about three letters between ⲁⲛⲁⲕ and ϩⲏⲧϥ. Instead of ϩⲣⲏⲓ ϩⲏⲧϥ (with T and sa 4, 5, 9), also non-elliptic ϯⲛ̅ϩⲏⲧϥ would be possible (cf. ϯϩⲙ̅ⲡⲓⲱⲧ sa 1).

---

1 ⲡⲓⲱⲧ] ⲡⲉⲓⲱⲧ rest. T 2 ⲙⲁⲥ rest.] ⲙⲙⲁⲥ rest. T | ϫⲉⲓⲟⲩⲁ] ϫⲉⲩⲁ T 2/3 ⲁⲛⲁⲕ ⲡⲉ] ⲁⲛⲁⲕ- T 3 ⲉϣⲱⲡⲉ] ⲉϣⲡⲉ T | ⲛⲃⲏⲩⲉ ⲉⲛ] em. ⲛⲛϩⲃⲏⲩⲉ ⲉⲛ; ⲉⲛ ⲛⲛϩⲃⲏⲩⲉ T | ⲡⲁ-] em. ⲙⲡⲁ- T 4 ⲉϣⲱⲡⲉ] ⲉϣⲡⲉ T | ϯⲉⲓⲣⲉ + ⲇⲉ corr. T | ⲙⲁⲩ] i.e. ⲙⲙⲁⲩ T 4/5 ⲉⲧⲉⲧⲛϣⲁⲧⲙ-] ⲉϣⲁⲧⲉⲧⲙ- (em. ⲉⲣϣⲁⲧⲉⲧⲛⲧⲙ-) T 5 ϩⲁⲛⲃⲏⲩⲉ] em. ⲁⲛϩⲃⲏⲩⲉ or ⲁⲛⲁϩⲃⲏⲩⲉ T 6 ⲉⲧⲉⲧⲛⲛⲁ-] ⲉⲧⲉⲧⲛⲁ- T | ϣⲟⲡ] ϣⲟⲟⲡ T | ϩⲏⲧ] i.e. ⲛϩⲏⲧ; ϩⲣⲏï ⲛϩⲏⲧ T (partly rest.) 7 ϩⲏⲧϥ rest.] i.e. ⲛϩⲏⲧϥ T | ϣⲓ] em. ϣⲓⲛⲉ T | ⲛⲥⲁⲙⲁϩⲧⲉ] em. ⲛⲥⲁⲁⲙⲁϩⲧⲉ; ⲥⲁⲁⲙⲁϩⲧⲉ T 8 ⲡ-] em. ⲙⲡ- T 9 ϩⲏⲧϥ] i.e. ⲛϩⲏⲧϥ T 10 ϭⲟⲩ rest.] ϭⲱ T 12 ⲛⲧⲁⲉⲓⲓⲱⲁⲛⲛⲏⲥ] i.e. ⲛⲧⲁïⲱⲁⲛⲛⲏⲥ T | ⲙⲓⲉ] ⲙⲏⲉ T 13 ⲛⲉⲩϩⲛ-] em. ⲛⲉⲩⲛ- T | ⲟⲩⲉⲉⲉ] em. ⲟⲩⲉⲉ T 14 ⲃⲏⲑⲁⲛⲓⲁⲁ] em. ⲃⲏⲑⲁⲛⲓⲁ T | ⲙⲁⲣⲓⲁⲙ] em. ⲙⲙⲁⲣⲓⲁⲙ; ⲙⲙⲁⲣⲓⲁ T 15 ⲧⲉ] ⲇⲉ T (em. ⲇⲉ ⲧⲉ or ⲧⲉ) | ⲛⲧⲁⲥⲧⲁϩⲁⲥ-] ⲛⲧⲁⲥⲧⲁϩⲥ- T 16 ⲁⲥϥⲛⲧ-] em. ⲁⲥϥⲁⲧ- T | ⲙⲡⲉⲥ-] ⲙⲡⲥ- T | ϥⲟⲩⲉ] ϥⲱⲉ T 16/17 ⲧⲉⲉⲓ ⲧⲉ] em. ⲧⲉⲉⲓ ⲉⲧⲉ-; ⲧⲏ ⲉⲧⲉ- rest. T 17 ⲡⲉⲥⲁⲛ] em. ⲡⲉⲥⲥⲁⲛ T | ϣⲁⲣⲁϥ] ϣⲁⲁⲣⲁϥ T (partly rest.) 18 ⲉⲓⲥϩⲏⲡⲉ] ⲉⲓⲥϩⲏⲧⲉ T | ⲉϥ-] (poss. em.) ϥ- rest. T 19 ⲡⲉⲧⲉⲕ-] ⲡⲉⲧⲕ- T (partly rest.) | ⲙⲁϥ] i.e. ⲙⲙⲁϥ T | ϭⲉ] ⲇⲉ T | ϭⲓ-] i.e. ⲛϭⲓ-

[ⲡ]ⲁϫⲉϥ ϫⲉ ⲡⲉⲉⲓϣⲱⲛⲉ ⲉϥϣⲟⲡⲡ ⲉⲛ ⲁⲡⲙⲟⲩ
[ⲁⲗ]ⲗⲁ ⲉⲧⲃⲉⲡⲉⲁⲩ ⲙⲡⲛⲟⲩⲧⲉ ϫⲉⲕⲁⲥⲉ ⲉⲣⲉⲡⲉϥ
[ϣ]ⲏ̣ⲣⲉ ⲛⲁϫⲓⲉⲁⲩ ⲁⲃⲁⲗ ϩⲓⲧⲟⲟⲧϥ ⲛⲉⲣⲉⲓ̅ⲥ̅ ⲟⲩⲱϣ 11:5
[ⲙ]ⲁⲣⲓⲁ ⲡⲉ ⲙⲛⲙⲁⲣⲑⲁ ⲧⲉⲥⲥⲱⲛⲉ ⲙⲛ̅ⲗⲁⲍⲁⲣⲟⲥ
[ⲛ]ⲧⲁⲣⲉϥⲥⲱⲧⲙ ⲇⲉ ϫⲉ ϥϣⲱⲛⲉ ⲧⲟⲧⲉ ⲙⲉⲛ ⲁⲩ 11:6
[ϭⲟ]ⲩ ⲛ̅ϩⲟⲟⲩ ⲥⲛⲉⲩ ⲛ̅ⲡⲙⲁ ⲉⲧⲉϥⲛ̅ϩⲏⲧϥ ⲙⲛ̅ⲥⲱⲥ 11:7
[ⲇ]ⲉ ⲡⲁϫⲉϥ ⲛⲉϥⲙⲁⲑⲏⲧⲏⲥ ϫⲉ ⲙⲁⲣⲁⲛ ⲁϯⲟⲩⲇⲁⲓⲁ
ⲡ̣ⲁϫⲉⲩ ⲛⲉϥ ⲛ̅ϭⲓⲙⲁⲑⲏⲧⲏⲥ ϫⲉ ⲣⲁⲃⲃⲓ ϯⲛⲟⲩ 11:8
[ⲛ]ⲉ̣ⲣⲉⲛⲓⲟⲩⲇⲁⲉⲓ ϣⲓⲛⲉ ⲥⲁⲛⲁϫⲱⲛⲉ ⲁⲣⲁⲕ ⲁⲩ
[ⲱ] ⲁⲛ ⲉⲕⲛⲁⲃⲱⲕ ⲁⲙⲉⲩ ⲁϥⲟⲩⲱϣⲃ ϭⲓⲓ̅ⲥ̅ 11:9
[ϫ]ⲉ̣ ⲙⲏ ⲙⲛⲙⲛ̅ⲥⲛⲁⲟⲩⲥ ⲛⲟⲩⲛⲟⲩ ϩⲛⲫⲟⲟⲩ
[ⲉⲣ]ⲉϣⲁⲟⲩⲉⲉ ⲙⲁϩⲉ ⲛ̅ⲫⲟⲟⲩ ⲙⲁϥϫⲓϫⲣⲁⲡ
[ϫ]ⲉ ⲛⲉϥⲛⲉⲩ ⲡⲉ ⲁⲡⲟⲩⲁⲉⲓⲛ ⲙ̅ⲡⲉⲉⲓⲕⲟⲥⲙⲟⲥ
[ⲉⲣ]ⲉϣⲁⲟⲩⲉⲉ ⲇⲉ ⲙⲁϩⲉ ϩⲛⲧⲟⲩϣⲓ ϣⲁⲩϫⲓϫⲣⲁⲡ 11:10
[ϫ]ⲉ ⲙⲛ̅ⲟⲩⲁⲉⲓⲛ ϣⲟⲡ ϩⲣⲏⲓ ϩⲏ̅ⲧ̅ⲥ̅ ⲛⲉⲉⲓ ⲁϥϫⲟⲟⲩ 11:11
[ⲁ]ⲩⲱ ⲙⲛⲥⲱⲥ ⲡⲁϫⲉⲩ ϫⲉ ⲗⲁⲍⲁⲣⲟⲥ ⲡϣⲃⲏⲣ
[ⲁ]ϥ̣ⲛ̅ⲕⲁⲧⲕ̣ⲉ ⲁⲗⲗⲁ ⲉⲓⲛⲁⲃⲱⲕ ϫⲉⲕⲁⲥⲉ ⲉⲓⲛⲁⲧⲟⲩ
ⲛ̣ⲟⲩⲛⲁⲥϥ ⲡⲁϫⲉϥ ϭⲉ ⲛⲉϥ ⲛ̅ϭⲓⲙⲁⲑⲏⲧⲏⲥ 11:12

11:7 26 In ⲡⲁϫⲉϥ, the letter ϥ is written over an oblique vertical stroke, very probably the left-hand part of an ⲩ. Thus the scribe was probably about, but had not yet finished, to write ⲡⲁϫⲉⲩ when he realized his mistake. He made the same mistake again in line 35 (and the opposite in line 37) without realizing it.

11:9 30 There is an empty space on the papyrus between ⲙⲛ̅ and ⲥⲛⲁⲟⲩⲥ but no trace of ink to suggest the presence of the lacking ⲧ.

---

20 ⲉϥ-] ϥ- T (prob. em. ⲉϥ-) | ϣⲟⲡⲡ] i.e. ϣⲟⲡ; ϣⲟⲟⲡ T 21 ⲡⲉϥ-] ⲡϥ- T 24 ⲁⲩ-] em. ⲁϥ- T 25 ϭⲟⲩ partly rest.] ϭⲱ T | ⲛ-] ϩⲛ- T | ⲉⲧⲉϥ-] ⲉⲧϥ- T | ⲙⲛⲥⲱⲥ] i.e. ⲙⲛⲛⲥⲱⲥ T 26 ⲛⲉϥ-] em. ⲛⲛⲉϥ- T 27 ⲙⲁⲑⲏⲧⲏⲥ] em. ⲙⲙⲁⲑⲏⲧⲏⲥ T 28 ⲥⲁ-] T; i.e. ⲛⲥⲁ- | ⲛⲁϫⲱⲛⲉ ⲁⲣⲁⲕ] ⲕⲁⲗⲉϩⲕ ⲙⲡⲱⲛⲉ T 29 ⲉⲕ-] ⲕ- T | ϭⲓ-] i.e. ⲛϭⲓ- 30 ⲙⲛⲥⲛⲁⲟⲩⲥ] em. ⲙⲛⲧⲥⲛⲁⲟⲩⲥ T 31 ⲙⲁϩⲉ] ⲙⲁⲁϩⲉ T | ⲛ-] ϩⲛ- T 33 ⲙⲁϩⲉ] ⲙⲁⲁϩⲉ T | ⲟⲩϣⲓ] ⲟⲩϣⲏ T | ϣⲁⲩ-] em. ϣⲁϥ-; ϣⲁⲣⲉϥ- rest. T 34 ϣⲟⲡ] ϣⲟⲟⲡ T | ϩⲏⲧⲥ] i.e. ⲛϩⲏⲧⲥ T 35 ⲙⲛⲥⲱⲥ] i.e. ⲙⲛⲛⲥⲱⲥ T | ⲡⲁϫⲉⲩ] em. ⲡⲁϫⲉϥ; ⲡⲁϫ̣ⲉϥ T | ⲡ-] ⲡⲛ- T 36 ⲉⲓⲛⲁ-^{2}] ⲉⲉⲓⲁ- T 36/37 ⲧⲟⲩⲛⲟⲩⲛⲁⲥϥ] em. ⲧⲟⲩⲛⲁⲥϥ T 37 ⲡⲁϫⲉϥ] em. ⲡⲁϫⲉⲩ; ⲡⲁϫ̣ⲉⲩ T | ⲙⲁⲑⲏⲧⲏⲥ] em. ⲙⲙⲁⲑⲏⲧⲏⲥ T

p. 7* John 11:12-28

ϫⲉ ⲡϫⲁⲉⲓⲥ ⲉϣⲱⲡⲉ ⲁϥⲛ̄ⲕⲁⲧⲕ̣[ⲉ ϥⲛⲁ]ⲧ̣ⲱⲱⲛ
ⲉⲣⲉⲓ̅ⲥ̅ ⲇⲉ ϫⲱ ⲙⲙⲁⲥ ⲉⲧⲃⲉⲡⲉ[ϥⲙⲟⲩ ⲛⲏ] ⲇⲉ ⲛⲉⲩ 11:13
ⲙⲉⲩⲉ ϫⲉ ⲉϥϫⲉⲡⲉⲉⲓⲛ̄ⲧⲁⲕⲉⲛ ⲛ̄[ⲧⲉⲡⲱⲃ]ⲉ̣ϣ ⲧⲟ 11:14
ⲧⲉ ⲁⲓ̅ⲥ̅ ϫⲟⲟϥ ⲛⲉⲩ ϩⲛⲛⲟⲩⲡⲁⲣⲏ[ⲥⲓⲁ ϫⲉ ⲗⲁ]ⲍⲁ
ⲣⲟⲥ ⲁϥⲙⲟⲩ ⲁⲩⲱ ϯⲣⲉϣⲉ ⲉⲧⲃⲉⲧⲏ̣[ⲛⲉ ϫⲉⲕ]ⲁ̣ⲥⲉ ⲉ 11:15
ⲣⲉⲧⲉⲧⲛⲣ̄ⲡⲓⲥⲧⲉⲩⲉ ϫⲉ ⲛⲉⲓⲙⲙ̣[ⲉⲩ ⲉⲛ ⲁⲗ]ⲗ̣ⲁ
ⲙⲁⲣⲁⲛ ϣⲁⲣⲁϥ ⲡⲁϫⲉⲑⲱⲙⲁⲥ [ⲡⲉϣⲁⲩⲙⲟ]ⲩ 11:16
ⲧⲉ ⲁⲣⲁϥ ϫⲉ ⲇⲓⲇⲩⲙⲟⲥ ⲛ̄ⲛⲉϥϣⲃⲣ̄[ⲙⲁⲑⲏⲧⲏⲥ ϫⲉ]
ⲙⲁⲣⲁⲛ ϩⲱⲟⲩⲛ ϫⲉⲕⲁⲥⲉ ⲉⲛⲁ[ⲙⲟⲩ ⲛⲙⲙⲉϥ]
ⲛ̄ⲧⲁⲣⲉϥⲓ ϭⲉ ϭⲓⲓ̅ⲥ̅ ⲁⲩϭⲓⲛⲉ ⲙⲁ[ϥ ⲉⲡⲉϥϥⲧⲁⲩ] 11:17
ⲡⲉ ϩⲛⲡⲧⲁⲫⲟⲥ ⲃⲏⲑⲁⲛⲓⲁ ⲇ[ⲉ ⲛⲉⲥϩⲏⲛ] 11:18
ⲁϩⲟⲩⲛ ⲁⲑⲓⲉⲣⲟⲩⲥⲟⲗⲏⲙ ⲛⲁⲙⲁⲃ ⲛ̣ⲥ̣[ⲧⲁ]
ⲇⲓⲟⲛ ⲟⲩⲙⲏϣⲉ ⲇⲉ ⲁⲃⲁⲗ ⲛⲓⲟⲩⲇⲁⲉⲓ ⲛⲉⲁ[ⲩ] 11:19
ⲉⲓ ϣⲁⲙⲁⲣⲑⲁ ⲙⲛⲙⲁⲣⲓⲁ ϫⲉ ⲉⲩⲛⲁⲥⲉⲗⲥ[ⲱⲗⲟⲩ]
ⲉⲧⲃⲉⲡⲟⲩⲥⲁⲛ ⲙⲁⲣⲑⲁ ⲇⲉ ⲛ̄ⲧⲁⲣⲉⲥⲥⲱⲧ[ⲙ] 11:20
ϫⲉ ⲓ̅ⲥ̅ ⲛⲏⲩ ⲁⲥⲓ ⲁⲁⲃⲁⲗ ϩⲏⲧϥ ⲙⲁⲣⲓⲁ ⲇⲉ
ⲛⲉⲥϩⲛⲙⲁⲥ ϩⲛⲡⲏⲉⲓ ⲡⲁϫⲉⲙⲁⲣⲑⲁ ϭⲉ 11:21
ⲛⲓⲓ̅ⲥ̅ ϫⲉ ⲡϫⲁⲉⲓⲥ ⲛⲉⲕϩⲛⲡⲉⲉⲓⲙⲁ ⲡⲉ
ⲛⲉⲣⲉⲡⲁⲥⲁⲛ ⲛⲁⲙⲟⲩ ⲉⲛ ⲁⲗⲗⲁ ϯⲛⲟⲩ 11:22
ⲁⲛ ϯⲥⲁⲩⲛⲉ ϫⲉ ϩⲱⲃ ⲛⲓⲙ ⲉⲧⲕⲁⲧⲁⲃϩ
ⲡⲛⲟⲩⲧⲉ ⲙⲁⲩ ⲉϥⲛⲁⲧⲉⲉⲓⲧⲟⲩ ⲛⲉⲕ
ⲡⲁϫⲉⲓ̅ⲥ̅ ⲛⲉⲥ ϫⲉ ⲡⲉⲥⲁⲛ ⲛⲁⲧⲱⲱⲛ 11:23
ⲡⲁϫⲉⲥ ⲛⲉϥ ϭⲓⲙⲁⲣⲑⲁ ϫⲉ ϯⲥⲁⲩⲛⲉ 11:24
ϫⲉ ϥⲛⲁⲧⲱⲱⲛ ϩⲛⲧⲁⲛⲁⲥⲧⲁⲥⲓⲥ ⲙ̅
ⲫⲁⲉ ϩⲛϩⲟⲟⲩ ⲡⲁϫⲉⲓ̅ⲥ̅ ⲛⲉⲥ ϫⲉ 11:25

1 ⲉϣⲱⲡⲉ] ⲉϣⲡⲉ T 2/3 ⲛⲉⲩⲙⲉⲩⲉ] ⲉⲩⲙⲉⲉⲩⲉ ⲡⲉ T, on account of ⲡⲉ probably to be emended (or restored, cf. Thompson's note) to ⲛⲉⲩ- (but cf. sa 10) 3 ⲉϥϫⲉ-] ⲉϥϫⲉⲣⲁ- T | ⲛⲧⲁⲕⲉⲛ] em. ⲛⲕⲁⲧⲕⲉ T | ⲱⲃⲉϣ] ⲱⲃϣ T 4 ϫⲟⲟϥ] ϫⲟⲟⲥ rest. T (cf. sa) | ϩⲛⲛ-] ϩⲛ- rest. T | ⲡⲁⲣⲏⲥⲓⲁ] ⲡⲁⲣⲣⲏⲥⲓⲁ T 5/6 ⲉⲣⲉⲧⲉⲧⲛ-] em. ⲉⲣⲉⲧⲉⲧⲛ(ⲛ)ⲁ-; ⲉⲣⲉⲧⲛⲁ- T 6 ⲛⲉⲓ-] ⲛⲉⲉⲓ- T 7 ϣⲁⲣⲁϥ] ϣⲁⲁⲣⲁϥ T | ⲡⲉϣⲁⲩ- rest.] ⲡⲉϣⲁⲣⲟⲩ- T 10 ϭⲓ-] i.e. ⲛϭⲓ- | ⲁⲩ-] ⲁϥ- T | ⲙⲁϥ] i.e. ⲙⲙⲁϥ rest. T 12 ⲑⲓⲉⲣⲟⲩⲥⲟⲗⲏⲙ] ⲑⲓⲉⲣⲟⲩⲥⲁⲗⲏⲙ T | ⲛⲁ-] ⲛⲛⲁ- T (partly rest.) | ⲙⲁⲃ] ⲙⲛⲧⲏ T 13 ⲛⲓⲟⲩⲇⲁⲉⲓ] em. ϩⲛⲛⲓⲟⲩⲇⲁⲉⲓ; ϩⲛⲛⲓⲟⲩⲧⲁⲉⲓ T 14 ϣⲁ-] ϣⲁⲁ- T | ⲉⲩⲛⲁ-] ⲉϥⲁ- T | ⲥⲉⲗⲥⲱⲗⲟⲩ] ⲥⲗⲥⲱⲗⲟⲩ T 16 ⲛⲏⲩ] ⲛⲛⲏⲩ T | ⲁⲁⲃⲁⲗ] em. ⲁⲃⲁⲗ rest. T 17 ϩⲛⲙⲁⲥ] ϩⲙⲁⲥⲧ T 18 ⲛⲓ-] em. ⲛ- T | ⲛⲉⲕϩⲛ-] i.e. ⲉⲛⲉⲕϩⲛ-; ⲛⲉⲕⲛ- T 20 ⲉⲧⲕⲁⲧⲁⲃϩ-] ⲉⲧⲕⲛⲁⲧⲃϩ- T 21 ⲙⲁⲩ] i.e. ⲙⲙⲁⲩ T | ⲉϥⲛⲁⲧⲉⲉⲓⲧⲟⲩ] ϥⲛⲁⲧⲉⲉⲧⲟⲩ T 23 ⲡⲁϫⲉⲥ ⲛⲉϥ ϭⲓⲙⲁⲣⲑⲁ] ⲡⲁϫⲉⲙⲁⲣⲑⲁ ⲛⲉϥ T 25 ϩⲛ-] em. ⲛ- T

ⲁⲛⲁⲕ ⲡⲉ ⲧⲁⲛⲁⲥⲧⲁⲥⲓⲥ ⲁⲩⲱ ⲡⲱϩ
ⲛⲉⲧⲣ̣ⲡⲓⲥⲧⲉⲩⲉ ⲁⲣⲁⲉⲓ ⲕⲁⲛⲛ ⲉϥ
ϣⲁⲙⲟⲩ ⲉϥⲛⲁⲱⲱϩ ⲁⲩⲱ ⲟⲩⲁⲛ 11:26
ⲛⲓⲙ ⲉⲧⲁⲁϩ ⲉⲧⲣⲡⲓⲥⲧⲉⲩⲉ ⲁⲣⲁⲉⲓ
ⲥⲉⲛⲁⲙⲟⲩ ⲉⲛ ϣⲁⲁⲛⲏϩⲉ ⲧⲉⲣⲡⲓⲥ
ⲧⲉⲩⲉ ⲁⲣⲁⲉⲓ ⲡⲁϫⲉⲥ ⲛⲉϥ ϫⲉ ⲥⲉ ⲡϫⲁⲉ̣ⲓⲥ 11:27
ⲁⲛⲁⲕ ⲧⲣⲡⲓⲥⲧⲉⲩⲉ ϫⲉ ⲛⲧⲁⲕ ⲡⲉ
ⲡⲉⲭ̄ⲣ̄ⲥ̄ ⲡϣⲏⲣⲉ ⲙⲡⲛⲟⲩⲧⲉ
ⲡⲉ̣ⲧ̣ⲛ̣ⲛⲏⲩ ⲁⲡⲕⲟⲥⲙⲟⲥ ⲁⲩⲱ 11:28

26 ⲱϩ] i.e. ⲱⲛϩ; ⲱⲱⲛϩ T 27 ⲛⲉⲧ-] em. ⲡⲉⲧ- T | ⲕⲁⲛⲛ] em. ⲕⲁⲛ T 28 ⲉϥⲛⲁ-] ϥⲛⲁ- rest. T | ⲱⲱϩ] i.e. ⲱⲛϩ; ⲱⲱⲛϩ T 29 ⲁⲁϩ] i.e. ⲁⲛϩ; ⲁⲁⲛϩ T 30 ⲥⲉⲛⲁ-] ϥⲛⲁ- T 32 ⲧⲣ-] em. ϯⲣ- T 33 ⲡⲉⲭ̄ⲣ̄ⲥ̄] ⲡ̄ⲭ̄ⲥ̄ T

p. 8* John 11:28-43

ⲧⲁⲣⲉⲥ[ⲭⲉⲛⲉⲉⲓ ⲁ]ⲥⲃⲱⲕ ⲁⲥⲙⲟⲩ̣ⲧ̣ⲉ̣ ⲁ̣ⲙ̣ⲁ̣ⲣⲓⲁ ⲧⲉⲥⲥⲱⲛ
ⲛⲉ ⲉⲥ[ⲭⲱ ⲙⲙⲁⲥ ⲛ]ⲉⲥ ⲭⲓⲟⲩⲉ ⲭⲉ ⲁⲡⲥϩ ⲉⲓ ⲁⲩⲱ ⲁϥ
ⲙⲟⲩⲧ[ⲉ ⲁⲣⲟ ⲛⲧ]ⲁⲥ̣ ⲇⲉ ⲛⲧⲁⲣⲉⲥⲥⲱⲧⲙ ⲥⲧⲱⲱⲛ ϩⲛ 11:29
ⲟⲩϭⲗ[ⲁⲙ ⲁⲥⲓ ⲁⲃ]ⲁⲗ ϣⲁⲁⲣⲁϥ ⲓ̅ⲥ̅ ⲇⲉ ⲛⲉⲡⲁⲧⲉϥⲓ ⲁϩⲟⲩⲛ 11:30
ⲁⲡϯ[ⲙⲉ ⲁⲗⲗⲁ ⲉ]ⲧⲓ ⲛⲉϥϩⲛⲡⲙⲁ ⲧⲁⲙⲁⲣⲑⲁ ⲧⲱⲙⲁⲧ̇
ⲁⲣⲁ[ϥ ϩⲏⲧϥ ⲛⲓⲟ]ⲩⲇⲁⲉⲓ ϭⲉ ⲇⲉ ⲉⲧⲙⲡⲏⲉⲓ ⲙⲙⲉⲥ ⲁⲩⲱ 11:31
ⲉⲧ[ⲥⲁⲗⲥⲗ ⲙⲙⲁⲥ] ⲛ̄ⲧⲁⲣⲟⲩⲛⲉⲩ ⲁⲙⲁ̣ⲣⲓⲁ ⲭⲉ ⲁⲥⲧⲱⲱⲛ
[ϩⲛⲟⲩϭⲗⲁⲙ] ⲁ̣ⲥⲓ ⲁⲃⲁⲗ ⲁⲩⲟⲩⲁϩⲟⲩ ⲛⲛ̄ⲥⲱⲥ ⲉⲩ
[ⲭⲱ ⲙⲙⲁⲥ ⲭ]ⲉ ⲉⲥⲁⲃⲱⲕ ⲁⲃⲁⲗ ⲁⲡⲧⲁⲫⲟⲥ ⲁⲣⲓⲙⲉ
[ⲙⲁⲣⲓⲁ ϭⲉ ⲛⲧ]ⲁ̣ⲣⲉⲥⲓ ⲁⲃⲁⲗ ⲁⲡⲙⲁ ⲉⲧⲉⲓ̅ⲥ̅ ϩⲏⲧϥ 11:32
[ⲁⲩⲱ ⲁⲥⲛⲉⲩ ⲁⲣ]ⲁ̣ϥ ⲁⲥⲛⲁⲭ ϩⲁⲛⲉϥⲟⲩⲣⲓⲧⲉ ⲉⲥⲭⲱ ⲙ
[ⲙⲁ]ⲥ̣ ⲛ̣ⲉ̣ϥ ⲭ̣[ⲉ] ⲡ̣ⲭⲁⲉⲓⲥ ⲛⲉⲕⲙⲡⲉⲉⲓⲙⲁ ⲡⲉ ⲛⲉⲣⲉ
ⲡ̣ⲁ̣ⲥⲁⲛ ⲛⲁⲙⲟⲩ ⲉⲛ ⲓ̅ⲥ̅ ϭⲉ ⲛ̄ⲧⲁⲣⲉϥⲛⲉⲩ ⲁⲣⲁⲥ ⲉⲥ 11:33
[ⲣⲓ]ⲙⲉ ⲁⲩⲱ ⲛⲓⲟⲩⲇⲁⲉⲓ ⲛ̄ⲧⲁⲩⲉⲓ ⲛⲙⲙⲉⲥ ⲉⲩⲣⲓⲙⲉ
[ⲁ]ⲩⲱ ⲁϥϣⲧⲁⲣⲧⲣ ϩⲛⲡⲉⲡ̅ⲛ̅ⲁ̅ⲁ ⲛⲑⲉ ⲛ̄ⲛⲉⲧⲙⲁⲭ ⲛ
[ϩ]ⲏⲧ ⲁⲩⲱ ⲡⲁⲭⲉϥ ⲭⲉⲉ ⲛⲧⲁⲧⲛⲕⲁ̣ⲁϥ ⲧⲟ ⲡⲁ 11:34
ⲭ̣ⲉⲩ ⲭⲉ ⲡⲭⲁⲉⲓⲥ ⲁⲙⲟⲩ ⲕⲛⲉⲩ ⲁϥⲣⲓⲙⲉ ⲛ̄ⲓ̅ⲥ̅ 11:35
[ⲛ]ⲉⲩⲭⲱ ⲇⲉ ⲙⲁⲥ ⲛ̄ϭⲓⲛⲓⲟⲩⲇⲁⲉⲓ ⲭⲉ̣ ⲛⲉⲩ ⲁⲑⲉ ⲉⲧ 11:36
ⲧ̣ϥⲙⲁⲉ ⲙⲁϥ ϩⲁⲉⲓⲛⲉ ϩⲏⲧⲟⲩ ⲡⲁ̣ⲭⲉⲩ ⲭⲉ ⲡⲉ 11:37
ⲉ̣ⲓ ⲧⲁϥⲟⲩⲉⲛ ⲛⲁⲃⲉⲗ ⲙⲡⲣ̄ⲃⲗⲗ̣ⲉ ⲛ̣ⲉ̣ⲙ̣ⲛϭⲁ̣ⲙ ⲙⲁϥ
ⲁⲧⲙⲧⲣⲉⲡⲉⲉⲓ ⲙⲟⲩ ⲓ̅ⲥ̅ ϭⲉ ⲁⲛ ⲛⲉϥⲙⲁⲕϩ ⲛ 11:38
ϩⲏⲧ̇ ϩⲣⲏⲓ ϩⲏⲧϥ ⲁϥⲓ ⲁⲡⲧⲁⲫⲟⲥ ⲛⲉⲩⲙϩⲉⲉⲩ ⲡⲉ

1 ⲧⲁⲣⲉⲥ-] ⲛⲧⲁⲣⲉⲥ- T 1/2 ⲧⲉⲥⲥⲱⲛ|ⲛⲉ] i.e. ⲧⲉⲥⲥⲱⲛⲉ (rest. T) 2 ⲭⲓⲟⲩⲉ] em. ⲛⲭⲓⲟⲩⲉ; ⲛϫⲓⲟⲩⲉ T | ⲥϩ] i.e. ⲥⲁϩ T | ⲁϥ-] ϥ- T 3 ⲥⲧⲱⲱⲛ] em. ⲁⲥⲧⲱⲱⲛ 4 ⲛⲉⲡⲁⲧⲉϥⲓ] i.e. ⲛⲉⲙⲡⲁⲧⲉϥⲓ T 5 ⲧⲁ-] i.e. ⲛⲧⲁ- T | ⲧⲱⲙⲁⲧ] ⲧⲱⲙⲧ T 6 ϩⲏⲧϥ rest. (lack of space)] ⲛϩⲏⲧϥ T | ⲇⲉ] del.; — T | ⲉⲧⲙ-] ⲉⲧϩⲛ- T | ⲙⲙⲉⲥ] em. ⲛ(ⲙ)ⲙⲉⲥ (rest. T) 8 ⲛⲛⲥⲱⲥ] em. ⲛⲥⲱⲥ T 10 ϩⲏⲧϥ] i.e. ⲛϩⲏⲧϥ T 11 ⲛⲁⲭ] em. ⲛⲁⲭⲥ; ⲛⲁϫⲥ T 12 ⲛⲉⲕ-] T; i.e. ⲉⲛⲉⲕ- | ⲡⲉ] — T 15 ϩⲛⲡⲉ-] ϩⲛⲡ- rest. T | ⲡ̅ⲛ̅ⲁ̅ⲁ] em. ⲡ̅ⲛ̅ⲁ̅ (rest. T) 16 ⲭⲉⲉ] em. ⲭⲉ; ϫⲉ T | ⲛⲧⲁⲧⲛ] ⲛⲧⲁⲧⲉⲧⲛ T 16/17 ⲡⲁⲭⲉⲩ+ⲛⲉϥ rest. T 17 ⲛ-] em. ⲛϭⲓ- 18 ⲙⲁⲥ] i.e. ⲙⲙⲁⲥ rest. T | ⲛⲉⲩ] ⲁⲛⲉⲩ T 18/19 ⲉⲧ|ⲧϥ-] i.e. ⲉⲧϥ- rest. T 19 ⲙⲁⲉ] em. ⲙⲁⲉⲓⲉ T | ⲙⲁϥ] i.e. ⲙⲙⲁϥ T | ϩⲁⲉⲓⲛⲉ+ⲇⲉ rest. T | ϩⲏⲧⲟⲩ] i.e. ⲛϩⲏⲧⲟⲩ rest. T 20 ⲧⲁϥ-] i.e. ⲛⲧⲁϥ- T (partly rest.) | ⲛⲁ-] em. ⲁⲛ- T | ⲙⲡⲣ-] em. ⲙⲡ-; ⲛⲡ- T | ⲙⲁϥ] i.e. ⲙⲙⲁϥ T; + ⲡⲉ T 21 ⲙⲁⲕϩ] ⲙⲁⲭ T (partly rest.) 22 ϩⲏⲧϥ] i.e. ⲛϩⲏⲧϥ T

ⲉⲩⲛⲟⲩⲱⲛⲉ ϩⲓⲣⲱϥ ⲡⲁϫⲉ̣ⲓ̅ⲥ̅ ϫⲉ ⲁ̣ⲗ̣ⲁϭⲉ ⲁ̣ⲡⲱ 11:39
ⲛⲉ ⲙⲙⲉⲩ ⲡⲁϫⲉⲥ ⲛⲉϥ ⲛ̄ϭⲓⲙⲁⲣ̣ⲑ̣ⲁ ϫⲉ ⲏⲇⲏ
ⲁϥⲣ̄ⲥⲧⲁⲉⲓ ⲡⲉϥϥ̄ⲧⲁⲩ ⲅⲁⲣ ⲡⲉ ⲡⲁϫⲉ̣ⲓ̅ⲥ̅ ⲛⲉⲥ 11:40
ϫⲉ ⲙⲡⲓϫⲟⲟⲥ ϫⲉ ⲉⲣⲉϣⲁⲣⲡⲓⲥⲧⲉⲩⲉ ⲧⲉⲛⲁⲛⲉⲩ
ⲁⲡⲉⲁⲩ ⲙ̄ⲡⲛⲟⲩⲧⲉ ⲁⲩϥⲓⲡⲱⲛⲉ ⲙⲙⲉⲩ ⲓ̅ⲥ̅ 11:41
ϭⲉ ⲁϥϥⲓⲉⲧϥ ⲁϩⲣⲏⲓ ⲁⲧⲡⲉ ⲉϥϫⲱ ⲙ̣ⲙ̣ⲁⲥ ϫⲉ
ⲡⲁⲉⲓⲱⲧ ϯϣⲡϩⲙⲁⲧ ⲧⲟⲟⲧⲕ ϫⲉ ⲁ̣ⲕⲥⲱ
ⲧⲙ ⲁⲣⲁⲉⲓ ⲁⲛⲁⲕ ⲇⲉ ϯⲥⲁⲩⲛⲉ ϫⲉ̣ [ⲕ]ⲥ̣ⲱⲧⲙ ⲁⲣ̣ⲁ̣ 11:42
ⲉⲓ ⲛ̄ⲟⲩⲁϣ ⲛⲓⲙ ⲁⲗⲗⲁ ⲧⲁⲉⲓϫⲟⲟϥ ⲉⲧ̣ⲃⲉⲡⲙⲏ
ϣⲉ ⲉⲧⲁϩⲉ ⲁⲣⲉⲧϥ ϫⲉⲕⲁⲥⲉ̣ ⲉϥⲛⲁⲣⲡⲓⲥ̣ⲧ̣ⲉⲩ
ⲉ ϫⲉ ⲛⲧⲁⲕ ⲡⲉ ⲛ̄ⲧⲁⲕ̣ⲧ̣ⲉ̣ⲩⲟⲩⲁ̣ⲉⲓ ⲁⲩⲱ 11:43
ⲧⲁⲣⲉⲓϫⲉⲛⲉⲉⲓ ⲁϥϣϭⲏⲗ̣ ⲁ̣ⲃ̣ⲁⲗ̣ ϩⲛⲛⲟⲩ

11:38 23 The letter ⲟ in ⲉⲩⲛⲟⲩⲱⲛⲉ is written over ⲱ. The scribe noticed his omission immediately, before writing ⲛⲉ.

11:42 33 In ⲛ̄ⲧⲁⲕ̣ⲧ̣ⲉ̣ⲩⲟⲩⲁ̣ⲉⲓ, only some traces of the upper parts of the letters ⲧⲉⲩ can be identified, and these are not unambiguous. On account of the suffix -ⲉⲓ, however, there can hardly be any doubt that the present MS did not use a form of the verb ⲧⲛⲛⲁⲩ (as did T and the Sahidic version) but rather a form of ⲧⲉⲩⲟ (cf. Bohairic).

---

24 ⲡⲁϫⲉⲥ ⲛⲉϥ ⲛϭⲓⲙⲁⲣⲑⲁ] ⲡⲁϫⲉⲙⲁⲣⲑⲁ ⲛⲉϥ T 28 ϥⲓⲉⲧϥ] i.e. ϥⲓ-ⲉⲓⲉⲧϥ; ϥⲓ-ⲛⲉϥⲃⲉⲗ T 29 ⲧⲟⲟⲧⲕ] i.e. ⲛⲧⲟⲟⲧⲕ T 31 ⲟⲩⲁϣ] em. ⲟⲩⲁⲉⲓϣ T | ⲧⲁⲉⲓ-] i.e. ⲛⲧⲁⲉⲓ- T 32 ⲉϥⲛⲁ-] ⲉⲩⲁ- T (partly rest.) 33 ⲧⲉⲩⲟⲩⲁⲉⲓ uncertain] ⲧⲛⲛⲁⲟⲩⲧ T 34 ⲧⲁⲣⲉⲓ-] em. ⲧⲁⲣⲉϥ-, i.e. ⲛⲧⲁⲣⲉϥ- T | ⲁϥϣϭⲏⲗ] em. ⲁϥⲁϣϭⲏⲗ T | ϩⲛⲛ-] ϩⲛ- T

p. 9* John 11:43-55

ⲛⲁϭ ⲥⲙⲏ ϫⲉ ⲗⲁⲍⲁⲣⲟⲥ ⲁⲙⲟⲩ ⲁⲃⲁⲗ ⲁϥⲓ ⲁⲃ̣ⲁⲗ ⲛϭⲓ 11:44
ⲡⲉⲛⲧⲁϥⲙⲟⲩ ⲉϥⲙⲏⲣ ⲁⲛⲉϥϭⲓϫ ⲙ[ⲛ]ⲛⲉϥ̣ⲟⲩⲣⲓ
ⲧⲉ ϩⲉⲛⲕⲉⲣⲉⲁ ⲁⲩⲱ ⲉⲡⲉϥϩⲟ ⲙⲏⲣ ϩⲛ̣[ⲟ]ⲩ̣[ⲥⲟ]ⲩ̣ⲇⲁ
ⲣⲓⲟⲛ ⲡⲁϫⲉⲓ̅ⲥ̅ ⲛⲉⲩ ϫⲉ ⲃⲁⲗⲉϥ ⲁⲃⲁⲗ [ⲧⲉⲧⲛⲕⲁⲁϥ]
ⲛϥⲃⲱⲕ ⲟⲩⲙⲏϣⲉ ⲇⲉ ⲁⲃⲁⲗ ϩⲛⲛⲓⲟⲩⲇⲁ̣ⲉⲓ ⲛ̄ⲧⲁ̣[ⲩⲉⲓ] 11:45
ϣⲁⲙⲁⲣⲓⲁ ⲁⲩⲱ ⲁⲩⲛⲉⲩ ⲁⲡⲉⲛⲧⲁϥⲉϥ ⲁⲩ[ⲣⲡⲓⲥ]
ⲧⲉⲩⲉ ⲁⲣⲁϥ ϩⲁⲉⲓⲛⲉ ⲇⲉ ⲁⲃⲁⲗ ϩⲏⲧⲟⲩ ⲁⲩⲃⲱⲕ ϣ[ⲁⲙ] 11:46
ⲫⲁⲣⲓⲥⲁⲓⲟⲥ ⲁⲩⲧⲁⲙⲁⲩ ⲁⲛⲉⲛⲧⲁⲓ̅ⲥ̅ ⲉⲧⲟⲩ ⲛⲁⲣⲭⲓ[ⲉ] 11:47
ⲣⲉⲩⲥ ⲇⲉ ⲙⲛⲛ̄ⲫⲁⲣⲓⲥⲁⲓⲟⲥ ⲁⲩⲥⲱ[ⲟⲩϩ ⲁϩ]ⲟⲩⲛ ϩⲛ
ⲟⲩⲥⲩⲛϩⲉⲇⲣⲩⲓⲟⲛ ⲡⲁϫⲉⲩ ϫⲉ ⲉⲩ [ⲡⲉ]ⲧ̣ⲛⲛⲁⲉϥ
ⲡⲉⲉⲓⲣⲱⲙⲉ ⲛⲉϥⲓⲣⲉ ⲛⲉⲉⲓ ⲙϩⲁ̣[ϩ ⲙⲙ]ⲁ̣ⲉ̣ⲓ̣ⲛ
ⲉⲛϣⲁⲕⲁⲁϥ ⲛ̄ⲧⲉⲉⲓϩⲉ ⲟⲩⲁⲛⲓⲙ ⲛ̣ⲁⲣ[ⲡⲓⲥⲧⲉⲩ] 11:48
ⲉ ⲁⲣⲁϥ ⲁⲩⲱ ⲛⲥⲉⲉⲓ ⲛ̄ϭⲓϩⲣⲱⲙⲁⲓⲟⲥ ⲛ̄ⲥⲉϥ̄ⲓ ⲧⲟ
ⲟⲧⲛ ⲙⲡⲉⲉⲓⲧⲟⲡⲟⲥ ⲁⲩⲱ ⲡⲛ̄ⲙⲏϣⲉ ⲟⲩⲉ ⲇⲉ 11:49
ⲁⲃⲁⲗ ϩⲏⲧⲟⲩ ϫⲉ ⲕⲁⲓⲫⲁⲥ ⲉⲡⲁ[ⲣ]ⲭ̣ⲓⲉⲣⲉⲩⲥ ⲡⲉ
ⲛ̄ⲧⲣⲁⲙⲡⲉ ⲉⲧⲛⲙⲉⲩ ⲡⲁϫⲉϥ ⲛ[ⲉ]ⲩ ϫⲉ ⲛⲧⲱ
ⲧⲛ ⲧⲉⲧⲛⲥⲁⲩⲛⲉ ⲉⲛ ⲛ̄ⲗⲁⲩⲉ ⲟⲩⲇⲉ ⲧⲉⲧⲛⲣⲗⲟ 11:50
ⲅⲓⲍⲉ ⲉⲛ ϫⲉ ⲥⲣ̄ⲛⲁϥⲣⲉⲉ ⲛⲉⲛ ϫⲉⲕⲁⲥⲉ ⲉⲣⲉ
ⲟⲩⲣⲱⲙⲉ ⲟⲩⲱⲧ ⲛⲁⲙⲟⲩ ϩⲁⲡⲗⲁⲟⲥ ⲛ̄ⲧⲉⲧⲛ
ⲡⲙⲏϣⲉ ⲧⲏⲣⲉϥ ϩⲁⲉⲓⲉ ⲁⲃⲁⲗ ⲛⲧⲁⲣⲉϥϫⲉ`ⲡ´ⲉⲉⲓ 11:51
ⲇⲉ ⲉⲛ ϩⲁⲣⲁϥ ⲟⲩⲁⲉⲧϥ ⲁⲗⲗⲁ ⲡⲁⲣⲭⲓⲉⲣⲉⲩⲥ ⲡⲉ
ⲛ̄ⲧⲣⲁⲙⲡⲉ ⲉⲧⲛⲙⲉⲩ ⲉϥⲣ̄ⲡⲣⲟⲫⲏⲧⲉⲩⲉ ϫⲉ

11:51 20 The scribe originally wrote ϫⲉⲛⲉⲉⲓ at the end of the line, and he corrected this by placing ⲡ above ⲛ. What he did with the letter ⲛ is not entirely clear; there is some blotting of ink, but it is difficult to decide whether this is due to cancellation or whether he tried to change ⲛ to ⲡ before writing ⲡ above the line.

---

1 ⲥⲙⲏ] em. ⲛⲥⲙⲏ T 2/3 ⲁⲛⲉϥϭⲓϫ ⲙⲛⲛⲉϥⲟⲩⲣⲓⲧⲉ] ⲛⲛⲉϥⲟⲩⲣⲓⲧⲉ ⲙⲛⲛⲉϥϭⲓϫ T 3 ϩⲉⲛ-] em. ⲛϩⲉⲛ- T (partly rest.) | ϩⲛ-] ⲛ- rest. T 4 ⲃⲁⲗⲉϥ] ⲃⲁⲗϥ T 6 ϣⲁ-] ϣⲁⲁ- T | ⲉϥ] rest. ⲉⲉϥ T 7 ϩⲏⲧⲟⲩ] i.e. ⲛϩⲏⲧⲟⲩ T 8 ⲉⲧⲟⲩ] ⲉⲉⲧⲟⲩ T 8/9 ⲁⲣⲭⲓⲉⲣⲉⲩⲥ] ⲁⲣϫⲓⲉⲣⲉⲩⲥ T 9 ⲙⲛⲛ-] ⲙⲛⲙ- T | ϩⲛ-] ⲛ- T 10 ⲟⲩ-] ⲡ- rest. T (cf. sa), but probably ⲟⲩ- to be restored | ⲥⲩⲛϩⲉⲇⲣⲩⲓⲟⲛ] ⲥⲩⲛϩⲉⲇⲣⲓⲟⲛ T | ⲡⲉⲧⲛⲛⲁⲉϥ] ⲡⲉⲧⲛⲁⲉⲉϥ T 11 ⲛⲉϥ-] ϥ- T | ⲛⲉⲉⲓ] em. ⲛⲛⲉⲉⲓ or del.; — T | ⲙϩⲁϩ] em. ⲛϩⲁϩ T 12 ⲟⲩⲁⲛⲓⲙ] i.e. ⲟⲩⲁⲛ ⲛⲓⲙ T 13 ϩⲣⲱⲙⲁⲓⲟⲥ] em. ⲛϩⲣⲱⲙⲁⲓⲟⲥ T 13/14 ⲧⲟⲟⲧⲛ] i.e. ⲛⲧⲟⲟⲧⲛ T 14 ⲧⲟⲡⲟⲥ] ⲙⲁ T | ⲙⲏϣⲉ] ϩⲉⲑⲛⲟⲥ T | ⲟⲩⲉ] ⲟⲩⲉⲉ T 15 ϩⲏⲧⲟⲩ] i.e. ⲛϩⲏⲧⲟⲩ T | ⲁⲣⲭⲓⲉⲣⲉⲩⲥ] ⲁⲣϫⲓⲉⲣⲉⲩⲥ T 16 ⲉⲧⲛⲙⲉⲩ] i.e. ⲉⲧⲙⲙⲉⲩ T 18 ⲛⲁϥⲣⲉⲉ] em. ⲛⲁϥⲣⲉ T 19 ⲟⲩⲱⲧ] ⲛⲟⲩⲱⲧ T | ⲛⲧⲉⲧⲛ-] em. ⲛⲧⲉⲧⲙ- T 20 ⲡⲙⲏϣⲉ] ⲫⲉⲑⲛⲟⲥ T | ⲧⲏⲣⲉϥ] ⲧⲏⲣϥ T | ⲛⲧⲁⲣⲉϥ-] em. ⲛⲧⲁϥ- T 21 ⲟⲩⲁⲉⲧϥ] ⲟⲩⲁⲉⲉⲧϥ T | ⲁⲣⲭⲓⲉⲣⲉⲩⲥ] ⲁⲣϫⲓⲉⲣⲉⲩⲥ T 22 ⲉⲧⲛⲙⲉⲩ] i.e. ⲉⲧⲙ-ⲙⲉⲩ T

ⲛⲉⲣⲉⲓ̅ⲥ̅ ⲛⲁⲙⲟⲩ ϩⲁⲡⲙⲏϣⲉ ⲁⲩⲱ ⲡⲙⲏ 11:52
ϣⲉ ⲟⲩⲁⲧϥ ⲉⲛ ⲁⲗⲗⲁ ⲭⲉⲕⲁⲥⲉ ⲁⲛ ⲛ̄ϣⲏⲣⲉ ⲧⲉ
ⲡⲛⲟⲩⲧⲉ ⲉⲧⲭⲁⲣⲉ ⲁⲃⲁⲗ ⲛϥⲥⲁϩⲟⲩ ⲁⲩⲙⲁ
ⲛ̄ⲟⲩⲱⲧ̇ ⲛ̄ⲫⲟⲟⲩ ϭⲉ ⲉⲧⲛⲙⲉⲩ ⲁⲩⲭⲓⲥⲁ 11:53
ⲭⲛⲉ ⲭⲉⲕⲁⲥⲉ ⲉⲩⲁⲙⲁⲟⲩⲧϥ ⲓ̅ⲥ̅ ϭⲉ ⲛⲉϥⲙⲁ 11:54
ϩⲉ ϭⲉ ⲉⲛ ⲡⲁⲣⲣⲏⲥⲓⲁ ϩⲛⲛⲓⲟⲩⲇⲁⲉⲓ ⲁⲗⲗⲁ
ⲁϥⲃⲱⲕ ⲁ̣ⲩⲭⲱⲣⲁ ⲉⲥϩⲏⲛ ⲁϩⲟⲩⲛ ⲁⲡⲭⲁⲓⲉ
ⲁⲩⲡⲟⲗⲓ ⲉⲩⲙⲟⲩⲧⲉ ⲁⲣⲁϥ ⲭⲉ ⲉⲫⲣⲁⲓⲙ ⲁϥϣⲱ
ⲡⲉ ⲙⲙⲉⲩ ⲙⲛⲛⲉⲩⲙⲁⲑⲏⲧⲏⲥ ⲛⲉⲩϩⲏⲛ 11:55
ⲇⲉ ⲁϩⲟⲩⲛ ϭⲓⲧⲡⲁⲥⲭⲁ ⲛⲓⲟⲩⲇⲁⲉⲓ ⲁⲩⲱ ⲁⲩ
ⲙⲏϣⲉ ⲃⲱⲕ ⲁϩⲣⲏⲓ ⲁⲑⲓⲉⲣⲟⲩⲥⲟⲗⲏⲙ
ⲧⲭⲱⲣⲁ ϩⲁⲑⲏ ⲙⲧⲡⲁⲥⲭⲁ ⲭⲉⲕⲁⲥⲉ ⲉⲩ

11:54 30 There is an empty space after ⲡⲟⲗⲓ, but no trace of ink which could suggest the former presence of a ⲥ. Judging from its length and uprightness, the sign after ⲡⲟⲗ should be the letter iota (not an abbreviation mark as in the common non-literary notation of this word).

11:55 34 There do not seem to be any letters written after ⲉⲩ at the end of the line, although there is another trace of ink (a dot?).

---

23 ⲡⲙⲏϣⲉ] ⲫⲉⲑⲛⲟⲥ T 23/24 ⲡⲙⲏϣⲉ] ⲫⲉⲑⲛⲟⲥ T 24 ⲟⲩⲁⲧϥ ⲉⲛ] lac. T (unclear, but perhaps to be restored as ⲟⲩ[ⲁⲉⲉⲧϥ ⲉⲛ]) | ⲧⲉ-] i.e. ⲛⲧⲉ- T 25 ⲭⲁⲣⲉ] ϫⲁⲣ rest. T, but probably ϫⲁⲣⲉ to be restored | ⲥⲁϩⲟⲩ] em. ⲥⲁⲩϩⲟⲩ T 26 ⲛⲫⲟⲟⲩ] ϩⲛⲫⲟⲟⲩ T | ⲉⲧⲛⲙⲉⲩ] i.e. ⲉⲧⲙⲙⲉⲩ T 27 ⲉⲩⲁ-] ⲉⲩⲛⲁ- T | ⲙⲁⲟⲩⲧϥ] ⲙⲟⲩⲟⲩⲧ ⲙⲙⲁϥ T 27/28 ⲙⲁϩⲉ] ⲙⲁⲁϩⲉ T 29 ⲭⲁⲓⲉ] ϫⲁⲉⲓⲉ T 30 ⲡⲟⲗⲓ] ⲡⲟⲗⲓⲥ T | ⲁⲣⲁϥ] ⲁⲣⲁⲥ T 31 ⲙⲛⲛⲉⲩ-] em. ⲙⲛⲛⲉϥ- T | ⲛⲉⲩ-] em. ⲛⲉⲥ-; ⲛⲉϥ- T 32 ϭⲓ-] i.e. ⲛϭⲓ- | ⲧⲡⲁⲥⲭⲁ] ⲡⲡⲁⲥⲭⲁ T | ⲛ-] em. ⲛⲛ- T 33 ⲑⲓⲉⲣⲟⲩⲥⲟⲗⲏⲙ] ⲑⲓⲉⲣⲟⲩⲥⲁⲗⲏⲙ T 34 ⲧⲭⲱⲣⲁ] em. (?) ϩⲛⲧⲭⲱⲣⲁ T | ⲙⲧⲡⲁⲥⲭⲁ] em. ⲛⲧⲡⲁⲥⲭⲁ; ⲙⲡⲡⲁⲥⲭⲁ T | ⲉⲩ] em. ⲉⲩⲁⲧⲟⲩⲃⲁⲩ T

p. 10* John 11:56-12:14

ⲛⲉⲩϣⲓⲛⲉ ϭⲉⲉ [ⲛ]ⲥⲁⲓ̅ⲥ̅ ⲁⲩⲱ ⲛⲉ̣ⲩϫⲱ ⲙⲁⲥ ⲛ̄ⲛⲟⲩ 11:56
ⲉⲣⲏⲩ ⲉⲩⲁϩⲉ [ⲁ]ⲣⲉⲧⲟⲩ ϩⲛⲡⲣⲡⲉⲉ ϫⲉ ⲉⲩ ⲡⲉⲧⲥⲇⲟϭⲓ ⲛⲏⲧⲛ
ϫⲉ ϥⲛⲏ̣ⲩ ⲉⲛ̣ ⲁ̣ⲡϣⲁⲉⲓⲉ ⲛⲁⲣⲭⲓⲉⲣⲉⲩⲥ ⲅⲁⲣ ⲙⲛⲙ̄ⲫⲁⲣⲓⲥⲁⲓⲟⲥ̣ 11:57
[ⲛⲉⲁⲩϯ ⲟⲩⲉⲛ]ⲧ̣ⲟⲗⲏ ϫⲉⲕⲁⲥⲉ ⲉⲣⲉⲟⲩⲉⲉ ⲙⲙⲉ ϫⲉ ⲉϥⲧⲟ ⲉϥ̣
[ⲁⲧ]ⲁ̣ⲙⲁⲩ ϫⲉⲕⲁⲥⲉ ⲉⲩⲛⲁϭⲁⲡϥ ⲓ̅ⲥ̅ ϭⲉ̣ ϩⲁⲑⲏ ⲛⲥⲁⲩ ⲛϩⲟⲟⲩ 12:1
[ⲁⲧⲡ]ⲁⲥⲭⲁ ⲁϥ̣ⲓ ⲁⲃⲏⲑⲁⲛⲓⲁ ⲁⲡⲙⲁ ⲉⲧⲉⲗⲁⲍⲁⲣⲟⲥ ⲛϩⲏⲧϥ ⲡⲉⲉⲓ
[ⲛⲧ]ⲁ̣ϥⲙⲟⲩ ⲁⲩ̣ⲱ ⲁⲓ̅ⲥ̅ ⲧⲟⲩⲛⲁⲥϥ ⲁⲃⲁⲗ ϩⲛⲛⲉⲧⲙⲁⲩⲟⲩⲉⲧ
[ⲁⲩ]ⲉ̣ⲓⲣⲉ ϭⲉ ⲛⲉϥ̣ ⲛⲟⲩⲇⲉⲓⲡⲛⲟⲛ ⲁⲡⲙⲁ ⲉⲧⲙⲉⲩ ⲁⲩⲱ ⲛⲉ 12:2
[ⲣⲉ]ⲙⲁⲣⲑⲁ ⲣ̣̅[ⲇⲓ]ⲁ̣ⲕ̣ⲟⲛⲁ ⲗⲁⲍⲁⲣⲟⲥ ⲇⲉ ⲛⲉⲟⲩⲉ ⲡⲉ ϩⲛⲛⲉⲧⲛⲏϫ
[ⲛ]ⲙ̣ⲉϥ ⲙⲁⲣ̣[ⲓⲁ ϭⲉ] ⲁ̣ⲥ̣ϫ̣ⲓ ⲛ̣ⲟⲩⲗⲓⲧⲣⲁ ⲛ̄ⲥⲁϭⲛ ⲛ̄ⲛⲁⲣⲇⲟⲥ ⲉⲥⲛ̄ 12:3
[ϩⲁⲧ] ⲉ̣ⲛⲛ[ⲁ]ϣ̣ⲉ̣ⲥⲟⲩⲛⲥⲧⲥ ⲁⲥⲧⲱϩⲥ ⲛ̄ⲛⲟⲩⲣⲓⲧⲉ ⲛⲓⲓ̅ⲥ̅ ⲁⲩⲱ
[ⲁⲥϥ]ⲱⲧⲉ ⲙⲙⲁ[ⲩ] ⲡⲉⲥ̣ⲟⲩⲉ ⲛⲧⲥⲁⲡⲉ ⲁⲡⲏⲉⲓ ⲇⲉ ⲙⲟ̣ⲩϩ ⲁⲃⲁⲗ
[ϩⲛ]ⲡⲥⲁⲧⲁⲉⲓ ⲙⲡ̣ⲥ̣ⲁϭⲛ ⲡⲁϫⲉⲓⲟⲩⲇⲁⲥ ⲡⲓⲥⲕⲁⲣⲓⲱⲧⲏⲥ 12:4
[ⲟⲩ]ⲉ̣ ⲁⲃⲁⲗ ϩⲛⲛ[ⲉϥ]ⲙⲁⲑⲏⲧⲏⲥ ⲡⲉⲉⲓ ⲉⲧⲛⲁⲣⲡⲁⲣⲁⲇⲓⲇⲟⲩ
[ⲙⲙ]ⲁ̣ϥ ϫⲉ ⲉⲧⲃⲉ̣[ⲉ]ⲩ ⲙⲡⲟⲩϯⲡⲉⲉⲓⲥⲁϭⲛ ⲁⲃⲁⲗ ϩⲁϣⲙ 12:5
[ⲧ]ϣⲱⲉ ⲛⲥⲧⲁⲧ̣ⲉ̣ⲣⲉ̣ ⲛ̄ⲥⲉⲧⲉⲧⲟⲩ ϩⲛϩⲏⲕⲉ ⲁⲗⲗⲁ ⲛⲉⲩⲥⲁⲛ 12:6
ϫ̣ⲓⲟⲩⲉ ⲡⲉ ⲁⲩⲱ ⲛⲉⲣⲉⲡⲅⲟⲗⲟⲥⲥⲟⲕⲟⲙⲟⲛ ⲛ̄ⲧⲟⲟⲧϥ

11:57 5 The letter ⲩ in ⲉⲩⲛⲁ- has almost the appearance of what is normally psi: a long vertical stroke with two short and bold oblique strokes extending from beneath the top end of the vertical stroke. The right-hand oblique stroke touches the following ⲛ; obviously there was no space for this stroke because the original letter was ϥ, and the correction was made only after other letters had been added. Note that the shape of secondary ⲩ here (with two diagonal arms) is different from that resulting from the correction on p. 5:25.

12:2 9 In ⲡⲉ ϩⲛ-, the middle stroke of ⲉ extends through the center of ϩ, thus giving the appearance of a letter ⳓ (which, however, is not used in this MS). Since this is not one of the usual ligatures of this scribe, one may assume that he first wrote ⲡⲉ ⲛ etc. (with the long middle stroke in ⲉ as sometimes found at the end of a word), and then inserted the ϩ as a correction.

---

1 ϭⲉⲉ] em. ϭⲉ T | ⲛⲥⲁ- rest.] ⲥⲁ- T | ⲙⲁⲥ] i.e. ⲙⲙⲁⲥ rest. T 3 ⲛⲏⲩ] ⲛⲛⲏⲩ T | ⲉⲛ+ⲁϩⲣⲏⲓ̈ T 4 ⲟⲩ- rest. (lack of space)] ⲛⲟⲩ- T | ⲉⲣⲉ-] ⲉⲣⲉϣⲁ- T 6 ⲧⲡⲁⲥⲭⲁ rest.] ⲡⲡⲁⲥⲭⲁ T 7 ⲙⲁⲩⲟⲩⲉⲧ] ⲙⲁⲟⲩⲧ T (partly rest.) 8 ⲇⲉⲓⲡⲛⲟⲛ] ⲇⲓⲡⲛⲟⲛ T (partly rest.) | ⲉⲧⲙⲉⲩ] i.e. ⲉⲧⲙⲙⲉⲩ T 9 ⲇⲓⲁⲕⲟⲛⲁ] ⲇⲓⲁⲕⲟⲛⲉⲓ T (partly rest.) | ⲛⲉⲟⲩⲉ] ⲛⲉⲩⲉⲉ T | ϩⲛ-] ⲛ- T 10 ⲛⲙⲉϥ rest.] ⲛⲙⲙⲉϥ T | ϭⲉ rest.] lac. T, to be restored as ϭⲉ rather than ⲇⲉ 11 ⲉⲛⲛⲁϣⲉ-] ⲉⲛⲁϣⲉ- T (partly rest.) | ⲥⲟⲩⲛⲥⲧⲥ] em. ⲥⲟⲩⲛⲧⲥ T | ⲛⲓ-] em. ⲛ- T 12 ⲡⲉⲥⲟⲩⲉ] em. ⲙⲡϥⲟⲩⲉ or (?) ⲙⲡⲉⲥϥⲟⲩⲉ; ⲙⲡϥⲱⲉ T 13 ⲥⲁⲧⲁⲉⲓ] em. (?) ⲥⲧⲁⲉⲓ T 14 ⲟⲩⲉ rest.] ⲟⲩⲉⲉ T (partly rest.) 16 -ϣⲱⲉ] em. -ϣⲉ T | ⲥⲧⲁⲧⲉⲣⲉ] ⲥⲧⲁⲧⲉⲉⲣⲉ rest. T | ⲧⲉⲧⲟⲩ] ⲧⲉⲉⲧⲟⲩ T | ϩⲛ-] em. ⲛⲛ- T (partly rest.) | ϩⲏⲕⲉ+ⲛⲧⲁϥϫⲉⲡⲉⲉⲓ ⲉⲛ ϫⲉ ⲡⲉϥⲣⲁⲟⲩϣ ⲡⲉ ⲉⲧⲃⲉⲛϩⲏⲕⲉ T (partly rest.)

ⲁ̣ⲩⲱ ⲛⲉϥϫⲓⲟ̣ⲩⲉ ⲛ̄ⲛⲉⲧⲟⲩⲛⲟⲩϫⲉ ⲙⲁⲩ ⲁⲣⲁϥ ⲡⲁϫⲉϥ 12:7
[ϭ]ⲉ ⲛϭⲓⲓ̅ⲥ̅ ϫⲉ ⲁⲗⲱⲧⲛ ϩⲁⲣⲁⲥ ϫⲉ ⲉⲥⲁⲣⲏϩ ⲁⲣⲁϥ ⲁⲫⲟⲟⲩ
[ⲧ]ⲁ̣ⲕⲉⲉⲥⲉ ⲛ̄ϩⲏⲕⲉ ⲛⲙⲙⲏⲧⲛ ⲛ̄ⲛⲟⲩⲁⲉⲓϣ ⲛⲓⲙ ⲁⲛⲁⲕ 12:8
[ⲇ]ⲉ ϯⲛⲙⲏⲧⲛ ⲉⲛ ⲛⲟⲩⲁⲉⲓϣ ⲛⲓⲙ ⲟⲩⲙⲏϣⲉ ⲇⲉ ⲉⲛⲛⲁϣϥ 12:9
[ⲁ]ⲃ̣ⲁⲗ ϩⲛⲛⲓⲟⲩⲇⲁⲉⲓ ⲁⲩⲥⲱⲧⲙ ϫⲉ ϥⲙⲙⲉⲩ ⲁⲩⲱ ⲁⲩⲉⲓ
ⲉⲧⲃⲉⲓ̅ⲥ̅ ⲉⲛ ⲟⲩⲁⲉⲧϥ ⲁⲗⲗⲁ ϫⲉⲕⲁⲥⲉ ⲉⲩⲁⲛⲉⲩ ⲗⲁⲍⲁ
ⲣⲟⲥ ⲡⲉⲛⲧⲁϥⲧⲟⲩⲛⲁⲥϥ ⲁⲃⲁⲗ ϩⲛⲛⲉⲧⲙⲁⲩⲟⲩⲉⲧ̇ ϫⲓ 12:10
ⲥⲁϫⲛⲉ ⲇⲉ ⲛ̄ϭⲓⲛⲁⲣⲭⲓⲉⲣⲉⲩⲥ ⲇⲉ ϫⲉⲕⲁⲥⲉ ⲉⲩⲁ
ⲙⲁⲟⲩⲧⲙⲡⲕⲉⲗⲁⲍⲁⲣⲟⲥ ϫⲉ ⲛⲉⲣⲉϩⲁϩ ⲛⲓⲟⲩⲇⲁⲉⲓ 12:11
ⲃⲏⲕ ⲡⲉ ⲉⲧⲃⲏⲧϥ ⲁⲩⲱ ⲛⲉⲩⲣ̄ⲡⲓⲥⲧⲉⲩⲉ ⲁⲓ̅ⲥ̅ ⲙ̅ 12:12
ⲡⲉϥⲣⲉⲥⲧⲉ ⲟⲩⲙⲏϣⲉ ⲉⲛⲁϣϥ ⲛⲉⲛⲧⲁⲩⲉⲓ ⲁⲡϣⲁ
ⲉⲓⲉ ⲛⲧⲁⲣⲟⲩⲥⲱⲧⲙ ϫⲉ ⲓ̅ⲥ̅ ⲛⲛⲏⲩ ⲁⲑⲓⲉⲣⲟⲩⲥⲟⲗⲏⲙ
ⲁⲩϫⲓ ⲛϩⲉⲛⲃⲁⲉ ⲁⲃⲁⲗ ϩⲛϩⲉⲛⲃ`ⲛ´ⲛⲉ ⲁⲩⲉⲓ ⲁⲃⲁⲗ 12:13
ϩⲓⲣⲉⲧϥ ⲁⲩⲱ ⲛⲉⲩⲁϣϭⲏⲗ ⲁⲃⲁⲗ ϫⲉ ϥⲥⲙⲁⲙⲁⲛ
ⲧ ⲛϭⲓⲡⲉⲧⲛⲛⲏⲩ ϩⲛⲡⲣⲉⲛ ⲡϫⲁⲉⲓⲥ ⲡⲣⲣⲟ
ⲙⲡⲓ̣ⲥ̣[ⲣⲁ]ⲏ̣ⲗ̣ ⲓ̣ⲥ̣̅ ϭⲉ ⲧⲁⲣⲉϥϭⲛ̣ⲟⲩⲉⲉⲓϣ ⲁϥⲧⲉⲗⲟ̣ 12:14

12:13 30 The scribe first wrote ϩⲛϩⲉⲛⲃⲛⲉ (with the word-final shape of ⲉ) and then added another ⲛ just above the last ⲛ.

---

18 ⲁⲩⲱ] — T | ⲙⲁⲩ] i.e. ⲙⲙⲁⲩ T 19 ⲉⲥⲁⲣⲏϩ] em. ⲉⲥⲁⲁⲣⲏϩ T 20 ⲧⲁ- rest.] em. ⲛⲧⲁ- T | ⲛⲛ-] i.e. ⲛ- T 21 ϯⲛⲙⲏⲧⲛ] i.e. ϯⲛⲙⲙⲏⲧⲛ T | ⲉⲛⲛⲁϣϥ] i.e. ⲉⲛⲁϣⲱϥ T 23 ⲟⲩⲁⲉⲧϥ] ⲟⲩⲁⲉⲉⲧϥ T | ϫⲉⲕⲁⲥⲉ + ⲁⲛ T 24 ⲙⲁⲩⲟⲩⲉⲧ] ⲙⲁⲟⲩⲧ T | ϫⲓ-] em. ⲁⲩϫⲓ-; ⲁⲩϫⲓ- T 25 ⲁⲣⲭⲓⲉⲣⲉⲩⲥ] ⲁⲣϫⲓⲉⲣⲉⲩⲥ T | ⲇⲉ[2]] del.; — T 26 ⲙⲁⲟⲩⲧⲙ-] em. ⲙⲁⲟⲩⲧ-; ⲙⲟⲩⲟⲩⲧ ⲛ- T | ⲛ-] ⲛⲛ- T (partly rest.) 28 ⲉⲛⲁϣϥ] i.e. ⲉⲛⲁϣⲱϥ T 29 ⲑⲓⲉⲣⲟⲩⲥⲟⲗⲏⲙ] ⲑⲓⲉⲣⲟⲩⲥⲁⲗⲏⲙ T 30 ϫⲓ ⲛ-] ϫⲓ- T 31/32 ⲥⲙⲁⲙⲁⲛⲧ] ⲥⲙⲁⲙⲁⲁⲛⲧ T 32 ⲡϫ.] em. ⲙⲡϫ.; ⲙⲡϫ. T 33 ϭⲉ] ⲇⲉ T | ⲧⲁⲣⲉϥ-] i.e. ⲛⲧⲁⲣⲉϥ- T | ⲟⲩⲉⲉⲓϣ] em. ⲟⲩⲉⲓⲱ T

p. 11* John 12:14-27

ⲁⲣⲁϥ·ⲕⲁⲧⲁⲑ[ⲉ ⲉⲧⲥ]ⲏ̣ϩ ϫⲉ ⲙⲡⲣ[ⲣϩⲁⲧⲉ] ⲧϣⲉⲉⲣⲉ̣ⲉ ⲛ 12:15
ⲥⲉⲓⲱⲛ ⲉⲓⲥⲡⲉ̣[ⲣⲣⲟ ⲛⲛ]ⲏⲩ ⲛ̣ⲉ ϥϩⲙⲁ̣[ⲥⲧ ⲁϫ]ⲛ̣ⲛⲟⲩⲥⲏϭ
ⲛ̣ⲉⲓⲱ ⲛⲉϥⲙⲁⲑ̣[ⲏⲧⲏ]ⲥ̣ ⲙ̅ⲡⲟⲩⲙⲙ̣[ⲉ ⲁⲛ]ⲉⲉⲓ ⲛ̅ϣⲁ̣[ⲣⲉⲡ] 12:16
[ⲁⲗⲗⲁ ⲧⲁⲣⲉϥ]ϫ̣ⲓⲉⲁⲩ ϭⲓ̣ⲓ̣̅ⲥ̣̅ ⲧⲟⲧⲉ [ⲁⲩⲣⲡⲙⲉⲩⲉ ϫⲉ ⲛⲉⲣⲉ]
[ⲛⲉⲉⲓ ⲥⲏϩ ⲉⲧ]ⲃ̣ⲏⲧϥ ⲁⲩⲱ ⲛⲉⲉⲓ ⲛ̣[ⲉⲛⲧⲁ]ⲩ̣ⲉ̣ⲧⲟⲩ ⲛ̣[ⲉϥ ⲁϥⲣ] 12:17
[ⲙⲛⲧⲣⲉ ϭⲉ] ϭⲓⲡⲙⲏϣⲉ ⲉⲧⲛ[ⲙⲙⲉ]ϥ̣ ϫⲉ ⲁϥ[ⲙⲟⲩⲧⲉ]
[ⲁⲗⲁⲍⲁⲣⲟⲥ] ⲁ̣ⲃⲁⲗ ϩⲛⲡⲧⲁⲫⲟⲥ [ⲁ]ⲩⲱ ⲁϥⲧⲟ[ⲩⲛⲁⲥ]ϥ̣ [ⲁ]
[ⲃⲁⲗ ϩⲛⲛⲉⲧⲙ]ⲁ̣ⲩⲟⲩⲉⲧ ⲉⲧⲃⲉⲡ[ⲉⲉⲓ ⲁ]ⲛ̣ ⲁⲡⲙⲏϣ̣ⲉ ⲉⲓ ⲁⲃⲁⲗ 12:18
ϩⲏ̣[ⲧ]ϥ̣ ϫ̣ⲉ̣ ⲁ̣ⲩ̣ⲥ̣ⲱ̣[ⲧ]ⲙ ϫⲉ ⲁϥⲣⲡⲉⲉⲓ[ⲙⲁⲉⲓ]ⲛ̣ ⲉⲙⲫⲁⲣ[ⲓⲥ]ⲁ̣ⲓⲟⲥ ⲇⲉ 12:19
ⲡⲁϫⲉⲩ ⲛⲟⲩⲉⲣⲏⲩ ϫⲉ ⲧⲉⲧⲉⲛⲛ̣[ⲉⲩ] ϫⲉ ⲧⲉⲧⲛϯ̣[ϩⲏ]ⲩ ⲉⲛ
ⲗⲁⲩⲉ ⲉⲓⲥⲡⲕⲟⲥⲙⲟⲥ ⲁϥⲃⲱⲕ ϩ̣[ⲓⲡⲁϩ]ⲟ̣ⲩ̣ ⲙⲁ̣ϥ̣ [ⲛ]ⲉ̣ⲩ̣[ⲛϩⲉ]ⲛ 12:20
ⲛⲟⲩⲁⲉⲓⲛⲓⲛ ⲁⲃⲁⲗ ϩⲛⲛⲉⲧⲃⲏⲕ [ⲁϩ]ⲣⲏⲓ ⲁⲡϣⲁⲉⲓⲉ ⲟⲩϣ̣
ϣⲧ ϩⲣⲏⲓ ϩⲛⲡϣⲁⲉⲓⲉ ⲉⲛ ⲛⲉⲉⲓ ϭ̣[ⲉ] ⲁⲩϯⲡⲟⲩⲁⲉⲓⲉ ⲁ 12:21
ⲫⲓⲗⲓⲡⲡⲟⲥ ⲁⲃⲁⲗ ϩⲛⲃⲏⲑⲥⲁⲓⲇⲁ̣ ⲛⲅⲁⲗⲓⲗⲁⲓⲁ ⲁⲩⲥⲉⲡ̣
ⲥⲱⲡϥ ⲉⲩϫⲱ ⲙⲁⲥ ϫⲉ ⲡϫⲁⲉⲓⲥ [ⲧ]ⲛⲟⲩϣ ⲁⲛⲉⲩ ⲁⲓ̅ⲥ̅
ⲁⲫⲓⲗⲓⲡⲡⲟⲥ ⲉⲓ ⲁϥϫⲟⲟⲥ ⲛⲁⲛⲇⲣⲉ̣ⲁ̣ⲥ̣ <ⲁⲛⲇⲣⲉⲁⲥ> ⲇⲉ ⲙⲛⲫⲓⲗⲓⲡ 12:22
ⲡⲟⲥ ⲁⲩϫⲟⲟⲥ ⲛ̅ⲥⲁⲓ̅ⲥ̅ ⲧⲁϥ ⲇⲉ̣ [ⲁϥⲟ]ⲩϣⲃ ⲉⲩϫⲱ ⲙ 12:23

12:18/19 9 The scribe's intention with regard to the redundant letter ⲉ between the two verses is not clear. The distance of this letter from the preceding ⲛ may suggest the word division given in the transcription, but the shape of the letter may suggest word-final ⲉ (thus possibly: ⲡⲉⲉⲓⲙⲁⲉⲓⲛⲉ ⲙⲫⲁⲣⲓⲥⲁⲓⲟⲥ). In either case the ⲉ would be unorthographic.

12:22 16 Different from the similar case of ⲛⲁⲉⲥⲁⲩ Jn 10:26/27, the word ⲛⲁⲛⲇⲣⲉⲁⲥ (haplography) is graphically well linked with both the preceding and following words.

---

1 ϣⲉⲉⲣⲉⲉ] em. ϣⲉⲉⲣⲉ rest. T 2 ⲥⲉⲓⲱⲛ] ⲥⲓⲱⲛ T | ϥ-] ⲉϥ- T | ⲁϫⲛⲛ-] ⲁϫⲛ- T | ⲥⲏϭ] ⲥⲓϭ rest. T, but probably ⲥⲏϭ to be restored 3 ⲛϣⲁⲣⲉⲡ] ϫⲓⲛϣⲁⲣⲉⲡ T (partly rest.) 4 ⲧⲁⲣⲉϥ- rest.] i.e. ⲛⲧⲁⲣⲉϥ- T | ϭⲓ-] i.e. ⲛϭⲓ- | ⲙⲉⲩⲉ rest.] ⲙⲉⲉⲩⲉ T 5 ⲉⲧⲟⲩ] ⲉⲉⲧⲟⲩ T 6 ϭⲓ-] i.e. ⲛϭⲓ- | ⲛⲙⲙⲉϥ rest.] ⲛⲙⲉϥ T 8 ⲙⲁⲩⲟⲩⲉⲧ] ⲙⲁⲟⲩⲧ T 9 ⲉⲙ-] em. ⲙ- T 10 ⲛⲟⲩ-] T; i.e. ⲛⲛⲟⲩ- | ⲧⲉⲧⲉⲛ-] ⲧⲉⲧⲛ- T | ⲧⲉⲧⲛ-] ⲛⲧⲛ- corr. T | ϯ-] ϭⲛ- T 11 ⲗⲁⲩⲉ] i.e. ⲛⲗⲁⲩⲉ T | ⲙⲁϥ] i.e. ⲙⲙⲁϥ T 11/12 ϩⲉⲛⲛ-] i.e. ϩⲉⲛ- T (partly rest.) 12 ⲟⲩⲁⲉⲓⲛⲓⲛ] ⲟⲩⲁⲉⲓⲁⲛⲓⲛ T (partly rest.) 12/13 ⲟⲩⲱϣⲧ] em. ⲁⲟⲩⲱϣⲧ; ⲁⲩⲱϣⲧ T 13 ϩⲣⲏⲓ ϩⲛ-] T to be restored thus | ⲉⲛ] del.; — T | ⲡⲟⲩⲁⲉⲓⲉ] T to be restored thus (i.e. ⲡⲟⲩ-ⲟⲩⲁⲉⲓⲉ) 14 ⲛⲅⲁⲗⲓⲗⲁⲓⲁ] em. ⲛⲧⲅⲁⲗⲓⲗⲁⲓⲁ T; + ⲁⲩⲱ T 15 ⲙⲁⲥ] i.e. ⲙⲙⲁⲥ T | ⲟⲩϣ] i.e. ⲟⲩⲱϣ T 16 ⲛⲁⲛⲇⲣⲉⲁⲥ] ⲁⲁⲛⲇⲣⲉⲁⲥ T (followed by ⲁⲛⲇⲣⲉⲁⲥ ⲇⲉ) 17 ⲛⲥⲁ-] ⲁ- T | ⲧⲁϥ] i.e. ⲛⲧⲁϥ T | ⲟⲩϣⲃ] i.e. ⲟⲩⲱϣⲃ T | ⲉⲩ-] em. ⲉϥ- T

ⲙⲁⲥ `ⲛ´ⲉⲩ ϫⲉ ⲁⲧⲟⲩⲛⲟⲩ ⲉⲓ ϫ̣ⲉ̣[ⲕⲁⲥⲉ] ⲉϥⲛⲁϫⲓ̣[ⲉ]
ⲁⲩ ⲛ̄ϭⲓⲡϣⲏⲣⲉ ⲡⲣⲱⲙⲉ ϩⲁⲙ̣[ⲏⲛ ϩⲁⲙⲏⲛ ϯϫⲱ] 12:24
ⲙⲁⲥ ⲛⲏⲧⲛ ϫⲉ ⲉⲣⲉⲧⲙⲧⲃⲓⲗⲃ[ⲓⲗⲉ ⲛⲥⲟⲩⲟ ϩⲁ]
ⲉⲓⲉ ⲁϩⲣⲏⲓ ⲁϫⲛⲡⲕⲁϩ ⲛ̄ⲥⲙⲟⲩ [ϣⲁⲥϭⲟⲩ ⲟⲩⲁⲉ]
ⲧ̄ⲥ̄ ⲉϣⲱⲡⲉ ⲇⲉ ⲉⲥϣⲁⲙⲟⲩ ϣⲁ[ⲥϯ ⲛⲟⲩⲕⲁⲣⲡⲟⲥ]
ⲉⲛⲛⲁϣϥ̄ ⲡⲉⲧⲙⲁⲉⲓⲉ ⲛⲧⲉϥⲯⲩ[ⲭⲏ ϥⲛⲁⲥⲁⲣⲙⲉⲥ] 12:25
ⲁⲩⲱ ⲡⲉⲧⲛⲁⲙⲙⲉⲥⲧⲱⲧⲉϥⲯⲩⲭ̣[ⲏ ϩⲛⲡⲉⲉⲓⲕⲟⲥ]
ⲙⲟⲥ ⲉϥϩⲛⲛⲁ ⲁϩⲣⲏⲓ ϩⲁⲣⲁⲥ [ⲁⲩⲱⲱϩ ϣⲁ]
ⲁⲛⲏϩⲉ ⲉϣⲱⲡⲉ ⲟⲩⲉⲉ ⲛⲁⲣⲇⲓ̣[ⲁⲕⲟⲛⲓ ⲙⲁⲣⲉϥ] 12:26
ⲟⲩϩ ⲥⲱⲉⲓ ⲁⲩⲱ ⲡⲙⲁ ⲁⲛⲁⲕ ⲉϯ̣[ⲛⲁⲃⲱⲕ ⲁⲣⲁϥ]
ⲡⲁⲕⲉⲇⲓⲁⲕⲟⲛⲟⲥ ⲛⲁϣⲱⲡⲉ [ⲙⲙⲉⲩ ⲉⲣⲉ]
ϣⲁⲟⲩⲉⲉ ⲣ̄ⲇⲓⲁⲕⲟⲛⲓ ⲛⲏⲓ ⲡⲁ[ⲉⲓⲱⲧ ⲛⲁⲧⲁⲉⲓⲁϥ]
ϯⲛⲟⲩ ⲁ̣ⲧⲁⲯⲩⲭⲏ ϣⲧⲁⲣⲧ̄ⲣ̄ ⲁⲩⲱ [ⲉⲩ ⲡⲉϯⲛⲁϫⲟⲟϥ] 12:27
ⲡ̣ⲁ̣ⲉ̣ⲓ̣ⲱ̣ⲧ̣ ⲛ̣ⲁⲧⲟⲩϫⲁⲉ̣[ⲓ] ⲁⲃⲁⲗ ϩⲛ̣[ⲧⲉⲉⲓⲟⲩⲛⲟⲩ ⲁⲗⲗⲁ ⲉⲧⲃⲉⲡⲉⲉⲓ]

12:23 18 The letters originally written in the line after ⲙⲙⲁⲥ are ϫⲉⲉⲩ (with ⲉⲩ in ligature). ⲛ has been added above the line just between the two ⲉ, and ϫⲉ is cancelled with a diagonal stroke.

12:25 25 All the letters representing the corrupt reading are well in line, there is no later addition or omission. (The corruption is not necessarily due to this scribe.)

---

18 ⲉϥⲛⲁ-] ⲉϥⲁ- T 19 ⲡⲣⲱⲙⲉ] em. ⲙⲡⲣⲱⲙⲉ T 19/20 ⲙⲁⲥ rest.] i.e. ⲙⲙⲁⲥ T 20 ⲃⲓⲗⲃⲓⲗⲉ] ⲃⲗⲃⲓⲗⲉ T 21 ϣⲁⲥ- rest.] ϣⲁⲣⲉⲥ- T 21/22 ⲟⲩⲁⲉⲧⲥ rest.] ⲟⲩⲁⲉⲉⲧⲥ T 22 ϣⲁⲥ- rest.] ϣⲁⲣⲉⲥ- T 23 ⲉⲛⲛⲁϣϥ] i.e. ⲉⲛⲁϣⲱϥ T | ⲥⲁⲣⲙⲉⲥ rest.] ⲥⲟⲣⲙⲉⲥ doubtful T 24 ⲛⲁⲙⲙⲉⲥⲧⲱ-] i.e. ⲛⲁⲙⲉⲥⲧⲱ- T 25 ⲉϥϩⲛⲛⲁ ⲁϩⲣⲏⲓ ϩⲁⲣⲁⲥ] i.e. ⲉϥⲛⲛⲁ etc.; ϥⲛⲁⲁⲣⲏϩ ⲁⲣⲁⲥ T | ⲱⲱϩ rest.] ⲱⲱⲛϩ T 26 ⲟⲩⲉⲉ] em. ⲟⲩⲛⲟⲩⲉⲉ T 27 ⲟⲩϩ] i.e. ⲟⲩⲱϩ; ⲟⲩⲁϩϥ T | ⲥⲱⲉⲓ] i.e. ⲛⲥⲱⲉⲓ T 29 ⲛⲏⲓ] ⲛⲏⲉⲓ T 31 ⲛⲁ-] ⲙⲁ- rest. T

p. 12* John 12:27-39

ⲁⲉ̣ⲓ̣ ⲁϩ[ⲣⲏⲓ . . .]ⲟⲩ ⲡⲁⲉⲓⲱ[ⲧ . . .].ⲛⲉ ⲙ̄ⲡⲉⲕ̣ 12:28
ⲣⲉ̣ⲛ̣ [ⲁⲩⲥⲙ]ⲏ ϭⲉ ⲉⲓ ⲁⲃⲁ[ⲗ ϩⲛ]ⲧ̣ⲡⲉ ϫⲉ ϯϯⲉ̣ⲁⲩ̣
[ⲁⲩ]ⲱ̣ [ⲁⲛ ϯⲛⲁ]ϯⲉⲁⲩ ⲡⲙⲏϣ̣[ⲉ] ⲇⲉ ⲉⲧⲁϩⲉ 12:29

12:27 1 From the alignment of extant margin on the lower left-hand fragment and the space needed for the restoration of all lower lines, it can be inferred with certainty that the top left-hand fragment is to be placed considerably nearer to the central part of the leaf than it is in the present conservation. (The same applies, of course, to the preceding verso page, p. 11, where the top right-hand fragment has to be moved to the left in order to be in agreement with the space required for restoring the lacunae in the first three lines.) This, however, creates problems for the restoration of the first line of p. 12: if ⲁϩ is part of ⲁϩⲣⲏⲓ, there is only room for three other letters. Even if ϯ- (instead of ⲧⲉⲉⲓ-) were assumed to be used as demonstrative article here, there would hardly be sufficient space for ⲁϯⲟⲩⲛⲟⲩ. Therefore, assuming a scribal error seems inevitable: he may have written ⲁϯⲛⲟⲩ (involving some misunderstanding of the text) or, in close sequence and ligature, ⲁⲧⲉⲉⲓⲟⲩ (note that the letters ⲛⲟⲩ are omitted by the scribe in another case: p. 11:26, and are duplicated in two cases: pp. 6:36/37 and 12:9).

12:28 1 The text around the second lacuna of that line, at the beginning of v 28, cannot be restored to what is expected: ⲡⲁⲉⲓⲱⲧ ϯⲉⲁⲩ ⲙ̄ⲡⲉⲕⲣⲉⲛ, because of the letters ⲛⲉ written before ⲙ̄ⲡⲉⲕⲣⲉⲛ (T has a lacuna here, too). The traces of ink immediately following the lacuna would suit an ⲩ so that ϯⲉⲁ]ⲩ might be restored; we would have to assume, then, that the scribe subsequently set out to copy a wrong text (e.g., ⲛⲉⲩ) but stopped after writing ⲛⲉ when he realized his mistake. In this case, however, one could expect that he would have cancelled the wrong letters ⲛⲉ. As he has not done so, it may seem more plausible to assume that these letters formed an integral part of what he was writing intentionally. Thus we might restore something like ϯⲥⲁ]ⲩ̣ⲛⲉ "My father, I know your name". Although this makes only poor sense, it may have resulted from misreading (in some previous exemplar) ϯⲉⲁⲩ for ϯⲥⲁⲩ and adding ⲛⲉ as a pseudocorrection, meant to obtain a real Coptic word. As there is a large lacuna in T, we cannot even be sure that T had a better reading.

---

1 ⲁⲉⲓ] i.e. ⲁⲓⲉⲓ; ⲁⲉⲓ rest. T | . . .ⲟⲩ] ⲁⲧⲉⲉⲓⲟⲩⲛⲟⲩ T (partly rest.); see note above | . . . .ⲛⲉ] lac. T; see note above 3 ⲇⲉ] — rest. T

[ⲁⲣⲉⲧ]ϥ̣ [ⲉ]ⲧ̣[ⲥⲱⲧ]ⲙ̣ ⲛⲉⲩϫⲱ ⲙⲁⲥ [ϫⲉ ⲟⲩϩⲣⲁⲩⲡⲉⲛ]
[ⲧⲁ]ϥϣⲱⲡ̣[ⲉ ϩⲉ]ⲛⲕⲁⲩⲉ ⲇⲉ ⲛⲉⲩ[ϫⲱ ⲙⲁⲥ ϫⲉ ⲟⲩ]
[ⲁⲅ]ⲅⲉⲗⲟⲥ ⲡ[ⲉⲛⲧⲁ]ϥⲥⲉϫⲉ ⲛⲙⲙ[ⲉϥ ⲁϥⲟⲩⲱϣⲃ] 12:30
[ⲉ]ϥ̣ϫⲱ ⲙⲙ[ⲁⲥ ϭ]ⲓⲓ̅ⲥ̅ ϫⲉ ⲛⲧⲁⲧ[ⲉⲉⲓⲥⲙⲏ ⲉⲓ ⲉⲛ]
[ⲉ]ⲧⲃⲏ̣ⲧ. ⲁ̣[ⲗⲗⲁ] ⲉⲧⲃⲉⲧⲏⲛⲉ ϯ[ⲛⲟⲩ ⲧⲕⲣⲓⲥⲓⲥ ⲙ] 12:31
ⲡ̣ⲕ̣ⲟⲥⲙ[ⲟⲥ ⲧ]ⲉ̣ ϯⲛⲟⲩⲛⲟⲩ ⲡ̣ⲁ̣ⲣⲭⲱⲛ ⲙ̣ⲡ̣[ⲉ]
[ⲉ]ⲓ̣ⲕ̣ⲟ̣ⲥⲙ̣[ⲟⲥ ⲥⲉⲛⲁⲛ]ⲁ̣ϫϥ ⲁⲃⲁⲗ ⲁⲛⲁⲕ `ϩ´ⲟⲩⲧ ⲛ̅ⲥⲉ 12:32
ϫ̣ⲉⲥⲧ ⲁϩ̣[ⲣⲏ]ⲓ ⲁϫⲛⲡⲕⲁϩ ⲛⲧⲁⲛⲥⲁⲕⲟⲩⲟⲩⲁⲛ
ⲛⲓ̣ⲙ ϣⲁ[ⲣⲁⲉ]ⲓ ⲛⲉϥϫⲱ ⲇⲉ ⲡⲉⲉⲓ ⲉϥⲣ̅ⲥ̅ⲏ`ⲙ̣´ⲁ̣ⲛ̣ⲛ̣ 12:33
ⲉ ϫⲉ ⲉϥⲛ̣[ⲁ]ⲙ̣ⲟ̣ⲩ ϩⲛⲛⲉϣ ⲙⲙⲟⲩ ⲁⲩⲟⲩϣⲃ ⲛⲉϥ 12:34
ϭⲓⲡⲙ[ⲏ]ϣ̣ⲉ ϫⲉ ⲁⲛⲁⲛ ⲁⲛ ⲁⲛⲥⲱⲧⲙ ⲁⲃⲁⲗ ϩⲛ

12:29 4/5 The letters surviving in line 5]ϥϣⲱⲡ̣[clearly suggest a restoration closer to the Sahidic version than what Thompson read (and restored) for T. What he meant to be there was probably ⲟⲩϩⲣⲁⲩ ⲡⲉ ⲛ[ⲧⲡⲉ] ⲛ̣ⲉⲉⲓ (at least one bracket is missing in his printed transcription, and the photograph does not provide much help here). If Thompson's reading is correct, T had a corrupt text here, and Thompson's suggestion to understand ⲡⲉⲉⲓ instead of ⲛ̣ⲉⲉⲓ would not cure it. If there were any chances to read something like ϣⲱⲡⲉ (after ⲡⲉⲛ[ⲧⲁϥ]?) in T, it would probably have been noticed by its editor. The present MS, on the other hand, does not provide enough space to restore the full text of the Sahidic version (which may be identical in this dialect: ⲟⲩϩⲣⲟⲩⲙⲡⲉ ⲡⲉⲛⲧⲁϥϣⲱⲡⲉ). The easiest explanation seems to be some sort of haplography — either ⲟⲩϩⲣⲟⲩⲙⲡⲉ <ⲡⲉ>ⲛⲧⲁϥϣⲱⲡⲉ, which could be a mistake directly made by the present scribe, or ⲟⲩϩⲣⲁⲩ ⲡⲉⲛⲧⲁϥϣⲱⲡⲉ as a result of "correcting" a previous error like, for instance, ⲟⲩϩⲣⲁⲩ {ⲛⲧ}ⲡⲉⲛⲧⲁϥϣⲱⲡⲉ. It still seems thinkable that the two Subachmimic MSS had a common base here.

12:32 10 The scribe originally wrote ⲁⲛⲁⲕⲟⲩⲧ and then inserted ϩ in small size half above the line, between ⲕ and ⲟ.

12:33 12 It is difficult to say what the scribe wrote after ⲉϥⲣ̅ⲥ̅ⲏ. Although the most plausible interpretation of the ink visible seems to be ⲁ̣ⲛ̣ⲛ̣ (and a trace of ink high above ⲏⲁ may represent a correction to ⲏⲙⲁ), something like ⲁ̣ⲓ̣ⲛ̣ⲛ̣ (or ⲁ̣ⲓ̣ⲛ̣ⲉ̣?) cannot be completely excluded.

---

4 ⲙⲁⲥ] i.e. ⲙⲙⲁⲥ T; + ⲡⲉ T 4/5 ⲟⲩϩⲣⲁⲩⲡⲉⲛⲧⲁϥϣⲱⲡⲉ rest.] i.e. ⲟⲩϩⲣⲁⲩ ⲡⲉⲛⲧⲁϥϣⲱⲡⲉ or (?) ⲟⲩϩⲣⲟⲩ(ⲙ)ⲡⲉ ⲛⲧⲁϥϣⲱⲡⲉ, em. (?) ⲟⲩϩⲣⲟⲩ(ⲙ)ⲡⲉ ⲡⲉⲛⲧⲁϥϣⲱⲡⲉ; T unclear, but perhaps to be restored as ⲟⲩϩⲣⲁⲩ ⲡⲉⲛⲧⲁϥϣⲱⲡⲉ 5 ⲙⲁⲥ rest.] i.e. ⲙⲙⲁⲥ rest. T 7 ϭⲓ- rest.] i.e. ⲛϭⲓ- 9 ϯⲛⲟⲩⲛⲟⲩ] em. ϯⲛⲟⲩ T 10 ϩⲟⲩⲧ] ϩⲱⲟⲩⲧ T | ⲛⲥⲉ-] ⲉⲩϣⲁ- T (partly rest.) 11 ⲁϫⲛ-] ϩⲓϫⲛ- rest. T (but perhaps ⲁϫⲛ- to be restored) | ⲛⲧⲁⲛ-] em. ⲛⲧⲁ-; ϯⲛⲁ- T | ⲥⲁⲕⲟⲩⲟⲩⲁⲛ] i.e. ⲥⲁⲕⲟⲩⲁⲛ T 12 ϣⲁⲣⲁⲉⲓ rest.] ϣⲁⲁⲣⲁⲉⲓ rest. T | ⲡⲉⲉⲓ] em. ⲙⲡⲉⲉⲓ T (partly rest.) 12/13 ⲥⲏⲙⲁⲛⲛⲉ (?)] i.e. ⲥⲏⲙⲁⲛⲉ rest. T 13 ϩⲛⲛ-] i.e. ϩⲛ-; T to be restored as ϩⲛⲉϣ ⲙⲙⲟⲩ | ⲁⲩ-] ⲁϥ- T | ⲟⲩϣⲃ] i.e. ⲟⲩⲱϣⲃ T 14 ϭⲓ-] i.e. ⲛϭⲓ- | ⲁⲛ] — T; probably del.

ⲡ̣ⲛⲟⲙ̣ⲟ̣ⲥ ϫⲉ ⲡⲉⲭ̅ⲥ̅ ⲛⲁϣⲱⲡⲉ ϣⲁⲁⲛⲏϩⲉ
[ⲁⲩ]ⲱ ⲛ̣ⲉ̣ϣ ϩⲉ ⲧⲁⲕ ϫⲱ ⲙⲙⲁⲥ ϫⲉ ⲥⲉⲛⲁϫⲉⲥⲧ
ⲡ̣[ϣⲏⲣ]ⲉ̣ ⲡⲣⲱⲙⲉ ⲡⲁϫⲉϥ ϭⲉ ⲛⲉϥ ⲛ̄ϭⲓ ⲓ̅ⲥ̅ 12:35
[ϫⲉ ⲉⲧⲓ] ⲕⲉⲕⲟⲩⲉⲓ ⲟⲩϣ ϣⲏⲙ ⲉⲡⲟⲩⲁⲉⲓⲛ ϣⲟⲡ
[ⲛⲙⲙⲏ]ⲧ̣ⲛ ⲙⲁϩⲉ ϩⲱⲥ ⲟⲩⲛⲧⲏⲧⲛ ⲡⲟⲩⲁⲉⲓⲛ
[ϫⲉ ⲛ]ⲉⲡⲕⲉⲕⲉ ⲧⲉϩⲁⲧⲏⲛⲉ ⲁⲩⲱ ⲡⲉⲧⲙⲁ
[ϩⲉ ϩ]ⲛ̣ⲡⲕⲉⲕⲉ ϥⲙⲙⲉ ⲉⲛ ϫⲉ ⲉϥϩⲛⲛⲁ ⲁⲧⲟ
[ⲟⲩⲛ]ⲧⲏⲧⲛ ⲡⲟⲩⲁⲉⲓⲛ ⲁⲣⲓⲡⲓⲥⲧⲉⲩⲉ ⲁⲡⲟⲩ 12:36
[ⲁⲉⲓⲛ] ϫⲉⲕⲁⲥⲉ ⲉⲣⲉⲧⲉⲧⲛⲛⲁϣⲱⲡⲉ ⲛ̣̄ϣⲏ
[ⲣⲉ ⲡⲟ]ⲩⲁⲉⲓⲛ ⲛ̣ⲉⲉⲓ ⲁϥϫⲟ̣ⲟⲩ ⲛϭⲓ ⲓ̅ⲥ̅ ⲁⲩⲱ
[ⲁϥϩⲁⲡ]ϥ ⲁⲣⲁⲩ ⲛⲉⲉⲓⲙⲁⲉⲓⲛ ⲇⲉ ⲧⲏⲣⲟⲩ ⲛ̣ⲧⲁϥ 12:37
[ⲉⲧⲟ]ⲩ ⲙⲡⲟⲩⲙⲡⲧⲟ ⲁⲃⲁⲗ ⲙ̄ⲡⲟⲩⲣ̄ⲡⲓⲥ
[ⲧⲉⲩⲉ] ⲁ̣ⲣⲁϥ ϫⲉⲕⲁⲥⲉ ⲉⲣⲉⲡⲥⲉϫⲉ ⲛⲏⲥⲁⲓⲟⲥ 12:38
[ⲛⲁ]ϫⲱⲕ ⲁⲃⲁⲗ ⲡⲉⲛ̣ⲧⲁϥϫⲟⲟϥ ϫⲉ ⲡ̣ϫⲁⲉⲓⲥ
[ⲛⲓⲙ] ⲡⲉ ⲛ̄ⲧⲁϥⲣ̄ⲡⲓⲥⲧⲉⲩⲉ ⲁⲡⲛϩⲣⲁⲩ ⲁⲩⲱ
[ⲡϭ]ⲃ̣ⲁ̣ⲉⲓ ⲙⲡϫⲁ̣ⲉⲓⲥ ⲛⲧⲁϥϭⲱⲗⲡ ⲁⲃ̣[ⲁⲗ]
[ⲛⲛⲓⲙ ⲉⲧⲃⲉⲡⲉⲉⲓ ⲙ]ⲡ̣ⲟⲩϭⲛϭⲁⲙ ⲁⲣ̣[ⲡⲓⲥⲧⲉⲩ] 12:39

12:36 22 The lacuna at the beginning of the line does not provide enough space for a conjunction to precede ⲟⲩⲛⲧⲏⲧⲛ (Thompson restores ϩⲱⲥ for T; sa has ϩⲟⲥⲟⲛ). But possibly the present MS had a circumstantial form: [ⲉⲩⲛ]ⲧⲏⲧⲛ.

---

15 ⲡⲉⲭ̅ⲥ̅] ⲡⲭ̅ⲥ̅ T 16 ϩⲉ] em. ⲛϩⲉ T | ⲧⲁⲕ] i.e. ⲛⲧⲁⲕ T | ϫⲱ] em. ⲕϫⲱ; ⲕϫⲱ T 17 ⲡⲣⲱⲙⲉ] em. ⲙⲡⲣⲱⲙⲉ T | ⲛⲉϥ] em. ⲛⲉⲩ T 18 ⲕⲉⲕⲟⲩⲉⲓ ⲟⲩϣ] em. ⲕⲉⲕⲟⲩⲉⲓ ⲛⲟⲩⲁⲉⲓϣ or ⲕⲉⲟⲩⲁⲉⲓϣ; ⲕⲉⲟⲩⲁⲉⲓϣ T | ϣⲏⲙ + ⲡⲉ T | ϣⲟⲡ] ϣⲟⲟⲡ T 19 ⲙⲁϩⲉ] ⲙⲁⲁϩⲉ corr. T | ⲡⲟⲩⲁⲉⲓⲛ] ⲙⲡⲟⲩⲁⲉⲓⲛ T 20/21 ⲙⲁϩⲉ rest.] ⲙⲁⲁϩⲉ T 21 ϩⲛⲛⲁ] i.e. ⲛⲛⲁ T | ⲁⲧⲟ + ϩⲱⲥ rest. T 22 ⲡⲟⲩⲁⲉⲓⲛ] ⲙⲡⲟⲩⲁⲉⲓⲛ T (partly rest.) 23 ⲉⲣⲉⲧⲉⲧⲛⲛⲁ] ⲉⲣⲉⲧⲛⲁ T 24 ⲡⲟⲩⲁⲉⲓⲛ rest.] em. ⲙⲡⲟⲩⲁⲉⲓⲛ T 24/25 ⲁⲩⲱ ⲁϥϩⲁⲡϥ ⲁⲣⲁⲩ ⲛⲉⲉⲓⲙⲁⲉⲓⲛ] thus to be restored in T 25 ⲛⲧⲁϥ-] ⲁϥ- rest. T, but space sufficient for ⲛⲧⲁϥ- 26 ⲉⲧⲟⲩ rest.] ⲉⲉⲧⲟⲩ rest. T | ⲙⲡⲧⲟ] i.e. ⲙⲧⲟ T 27 ⲏⲥⲁⲓⲟⲥ] ⲏⲥⲁϊⲁⲥ rest. T 28 ⲛⲁ- rest.] ⲁ- rest. T 30 ⲙ-] ⲛ- T

p. 13* John 12:39-13:2

[ⲉ ϫⲉ ⲁϥϫⲟⲟⲥ ⲁⲛ ⲛϭⲓⲏⲥ]ⲁ̣ⲓⲟⲥ ϫ̣[ⲉ ⲁ]ϥⲧⲱⲙ 12:40
[ⲛⲛⲟⲩⲃⲉⲗ ⲁⲩⲱ ⲁϥⲧⲱⲙ ⲙⲡ]ⲟ̣ⲩϩⲏⲧ [ϫ]ⲉⲕⲁⲥⲉ
[ⲛⲟⲩⲛⲉⲩ ϩⲛⲛⲟⲩⲃⲉⲗ ⲛⲥⲉⲣⲛⲟⲉ]ⲓ ϩⲛⲡⲟⲩ̣ϩ̣ⲏⲧ ⲛ
[ⲥⲉⲕⲁⲧⲟⲩ ⲧⲁⲧⲁⲛϩⲁⲩ .. ⲁϥ]ϫⲟⲟⲥ ⲛ̄ϭⲓⲏⲥⲁⲓⲟⲥ 12:41
[ϫⲉ ⲁϥⲛⲉⲩ ⲁⲡⲉⲁⲩ ⲙⲡⲛⲟⲩⲧⲉ] ⲁ̣ⲩⲱ ⲁϥⲥⲉϫⲉ ⲉⲧⲃ[ⲏ]
ⲧ̣ϥ ⲁ̣ⲃ̣[ⲁⲗ] ϩ[ⲛⲛⲁ]ⲣ̣[ϫⲱⲛ ⲁϩⲁϩ] ⲣⲡⲓⲥⲧⲉⲩⲉ ⲁⲣⲁϥ ⲁⲗⲗ[ⲁ] 12:42
ⲉⲧⲃⲉⲙⲫⲁⲣⲓⲥⲁⲓⲟⲥ ⲛⲉⲩ[ⲣϩⲟⲙⲟ]ⲗ̣ⲟⲅⲉⲓ ⲉⲛ ϫⲉⲕⲁⲥⲉ
ⲛⲟⲩϣⲱⲡⲉ̣ [ⲛⲁⲡⲟⲥⲩⲛⲁⲅⲱⲅ]ⲟ̣ⲥ ⲁⲩⲙⲣ̄ⲣⲉⲡⲉ 12:43
ⲁⲩ ⲅⲁⲣ ⲛⲣⲱⲙ̣[ⲉ ⲛϩⲟⲩⲟ ⲁⲡⲉⲁⲩ] ⲙⲡⲛⲟⲩⲧⲉ ⲓ̄ⲥ̄ ⲇⲉ ⲁϥ[ⲁ] 12:44
ϣϭⲏⲗ ⲁⲃⲁⲗ̣ [ⲉϥϫⲱ ⲙⲙⲁⲥ ϫⲉ] ⲡ̣ⲉⲧⲣⲡⲓⲥⲧⲉⲩⲉ ⲁⲣⲁ̣[ⲉⲓ]
ⲉϥⲣⲡⲓⲥⲧⲉⲩⲉ [ⲁⲣⲁⲉⲓ ⲉⲛ ⲁⲗⲗⲁ ⲡⲉⲛⲧⲁϥⲧⲉ]ⲩ̣ⲟ̣ⲩ[ⲁ]ⲉ[ⲓ] ⲁ̣ⲛ̣[ⲁⲕ] 12:46
ⲡⲉ ⲡⲟⲩⲁⲉⲓⲛ ⲛ̣ⲧ̣[ⲁ]ⲉ̣[ⲓ]ⲉ̣ⲓ̣ ⲁ̣[ⲡⲕⲟⲥⲙⲟⲥ ϫⲉⲕⲁⲥⲉ ⲟⲩⲁⲛ]
ⲛⲓⲙ ⲉⲧⲣⲡⲓⲥⲧⲉⲩⲉ ⲁⲣⲁⲉⲓ ⲛⲟ[ⲩϭⲱ ϩⲛⲡⲕⲉⲕⲉ ⲁⲩⲱ] 12:47
ⲉⲣⲉϣⲁⲟⲩⲉⲉⲉ ⲥⲱⲧⲙ ⲁⲛⲁⲥⲉ̣[ϫⲉ ⲛϥⲧⲙⲁⲣⲏϩ ⲁⲣⲁⲩ]

12:40 3 The restored wording given in the transcription is probably too long for the lacuna, although the MS reading may have been essentially the same, perhaps involving some reductions: ϩⲛⲟⲩⲃⲉⲗ and/or ⲥⲉⲣ̄ⲛⲟⲉⲓ.

12:41 4 This MS cannot have started the sentence with ⲛⲉⲉⲓ or any other demonstrative pronoun, since it has ⲁϥϫⲟⲟⲥ instead of ⲁϥϫⲟⲟⲩ (T with sa). Perhaps it used a Second Perfect: ⲛⲧⲁϥ]ϫⲟⲟⲥ, or there was an empty space between the two sentences.

12:47 14 What has been transcribed as ⲟⲩⲉⲉⲉ ⲥⲱⲧⲙ does not make easy reading in the MS. The second ⲉ has the word-final shape (long middle-stroke), and what follows it may be read as either ⲉⲧⲥⲱⲧⲙ, ⲉⲥⲥⲱⲧⲙ, or ⲉⲥⲱⲧⲙ. In the latter case, a correction may be involved, imposing the shape of a long ⲥ (with an extended upper horizontal stroke) on what may previously have been ⲧⲥ or something else. — In the second half of the line, it is not quite clear whether ⲛϥ(ⲧⲙ)ⲁⲣⲏϩ should be restored with or without negation (T also has a lacuna here). Judging from the amount of space available in the lacunae, it seems almost certain that T did not have the negation (along with sa), whereas in the present MS negative ⲧⲙ- would well fit the lacuna. If it indeed had ⲧⲙ-, this would be the only MS known to have the Conditional construction in 12:47 (*S* ⲉⲣϣⲁⲛⲟⲩⲁ ⲥⲱⲧⲙ) followed by a negated verb "keep". So far all Coptic versions with a negative Conjunctive (rendering Greek καὶ μὴ φυλάξῃ), i.e., mf and bo, start the sentence with a ⲡⲉⲧⲛⲁⲥⲱⲧⲙ construction (ⲡⲉⲧⲛⲉ- *W*, ⲫⲏ ⲉⲧⲛⲁ- *B*), while the Conditional construction is found in those versions that render καὶ φυλάξῃ, i.e., sa (total, *pace* Horner) pbo. If this concurrence is more than incidental, negative ⲧⲙ- would seem less probable for the present MS.

---

1 ⲏⲥⲁⲓⲟⲥ] ⲏⲥⲁⲓ̈ⲁⲥ T 4 ..ⲁϥϫⲟⲟⲥ] poss. restore ⲛⲧⲁϥϫⲟⲟⲥ; ⲛⲉⲉⲓ ⲁϥϫⲟⲟⲩ T (partly rest.) | ⲏⲥⲁⲓⲟⲥ] ⲏⲥⲁⲓ̈ⲁⲥ rest. T 7 ϩⲟⲙⲟⲗⲟⲅⲉⲓ] ϩⲟⲙⲟⲗⲟⲅⲓ T 9 ⲛⲣⲱⲙⲉ] i.e. ⲛⲛⲣⲱⲙⲉ T 11 ⲧⲉⲩⲟⲩⲁⲉⲓ rest.] ⲧⲉⲩⲁⲉⲓ T 12 ⲛⲧⲁⲉⲓⲉⲓ rest.] ⲛⲧⲁⲉⲓ T 13 ⲛⲟⲩ-] ⲛⲉϥ- rest. T (cf. Jn 11:26) 14 ⲟⲩⲉⲉⲉ] em. ⲟⲩⲉⲉ rest. T | ⲛϥⲧⲙⲁⲣⲏϩ rest.] or better (?) ⲛϥⲁⲣⲏϩ rest. T

ⲁⲛⲁⲕ ϯⲛⲁⲣⲕⲣⲓⲛⲉ ⲙⲁϥ ⲉⲛ ⲛ̣[ⲧⲁⲉⲓ ⲅⲁⲣ ⲉⲛ ⲁⲣ]
ⲕⲣⲓⲛⲉ ⲙⲡⲕⲟⲥⲙⲟⲥ ⲁⲗⲗⲁ ϫ[ⲉⲕⲁⲥⲉ ⲉⲓⲛⲁⲛⲁϩⲙϥ]
ⲡⲉⲧⲣ̣ⲁⲑ̣ⲉⲧⲓ ⲙⲁⲉⲓ ⲉϥϫⲓ ⲉⲛ ⲛ̣[ⲛⲁⲥⲉϫⲉ] ⲟ̣[ⲩⲛⲧⲉϥ] 12:48
ⲙⲡⲉ̣ⲧⲛⲛⲁ̣ⲣⲕⲣⲓⲛⲉ ⲙⲁϥ ⲡⲥ̣[ⲉϫⲉ ⲛⲧⲁ]ⲉ̣ⲓϫⲟⲟ̣[ϥ]
ⲛ̄ⲧⲁϥ ⲉⲧⲛⲛⲁⲣⲕⲣⲓⲛⲉ̣ ⲙⲁϥ [ϩⲛⲫⲁⲉ ⲛ]ϩⲟⲟⲩ
ϫⲉ ⲁⲛ̣ⲁⲕ ⲉⲛ ⲡⲉⲛⲧⲁⲉⲓⲥⲉ[ϫⲉ ϩⲁⲣⲁ]ⲉ̣[ⲓ] ⲟ̣ⲩ[ⲁⲉⲧ] 12:49
ⲁⲗⲗⲁ ⲡⲁⲉⲓⲱⲧ ⲛⲧⲁϥⲧⲉⲩⲟⲩⲁ̣[ⲉⲓ] ⲡⲉⲛⲧⲁϥϯ [ⲛⲏⲓ]
ⲛⲟⲩⲉ̣ⲛⲧⲟⲗⲏ ϫⲉ ⲉⲩ ⲡ̣ⲉⲧⲛⲛ̣[ⲁ]ϫⲟⲟϥ ⲁⲩⲱ [ⲉⲩ ⲡⲉ]
ϯⲛⲁⲧⲉⲩⲁ̣ϥ ⲁⲩⲱ ϯⲥⲁⲩⲛⲉ̣ ϫⲉ [ⲧϥ]ⲉⲛⲧⲟⲗⲏ ⲟⲩⲱ[ⲱϩ] 12:50
ⲧⲉ ⲛϣⲁⲁⲛⲏϩⲉ ⲛⲉϯϫⲱ ⲙⲁ[ⲩ] ⲁⲛⲁⲕ ⲕⲁⲧⲁⲑ̣ⲉ̣ [ⲛ]
ⲧⲁⲡⲁⲉⲓⲱⲧ ϫⲟⲟⲥ ⲛⲏⲓ ⲧⲉⲉ̣[ⲓ ⲧⲉ ⲑⲉ] ⲉ̣ϯⲥⲉϫⲉ ⲙ
ⲙⲁⲥ ϩⲁ̣ⲑⲏ ⲙⲡϣⲁⲉⲓⲉ ⲛⲧ̣[ⲡⲁⲥ]ϫⲁ̣ [ⲉϥⲥⲁⲩ] 13:1
ⲛⲉ ⲛ̄ϭⲓⲓ̅ⲥ̅ ϫⲉ ⲁ̣[ⲧϥⲟⲩⲛⲟⲩ ⲉⲓ ϫⲉ ⲉϥⲁⲡⲱⲛⲉ ⲁⲃⲁⲗ]
ϩⲛⲡⲉⲉⲓⲕ[ⲟⲥⲙⲟⲥ ⲛϥⲃⲱⲕ ϣⲁⲡⲉⲓⲱⲧ ⲁϥ]
ⲙⲣⲣⲉ̣ⲛ̣[ⲉⲧⲉ ⲛⲱϥ ⲛⲉ ⲉⲧϩⲛⲡⲕⲟⲥⲙⲟⲥ]
[ⲁϥⲙⲣⲣⲓⲧⲟⲩ ϣⲁⲁⲃⲁⲗ ⲁⲩⲱ ⲛⲧⲁⲣⲉ] 13:2

12:47 16 The form ⲉⲓⲛⲁⲛⲁϩⲙϥ has been restored with a view to the scribe's apparently regular spelling ⲉⲓ- (for ⲉⲉⲓ- or ⲉϊ-) in connection with ⲛⲁ-. As regards the last syllable of this word, there is no attestation for the spelling of stem-final /CN⸗/ plus consonantal suffix in this MS. In view of the scribe's habit of sonorant cluster resolution in some other cases (see *supra*, Introduction), the spelling may have been ⲛⲁϩⲙⲉϥ. In T, however, it seems clear that ⲉⲉⲓⲛⲁⲛⲁϩ̅ⲙ̅ϥ̅ should be restored (cf. ⲟⲩⲁϩ̅ⲙ̅ϥ̅ Jn 9:7 and ⲥⲁⲧ̅ⲙ̅ϥ̅ Jn 16:13).

---

15 ⲙⲁϥ] i.e. ⲙⲙⲁϥ T 17 ⲡⲉⲧⲣⲁⲑⲉⲧⲓ] ⲡⲉⲧⲁⲑⲉⲧⲓ rest. T | ⲙⲁⲉⲓ] i.e. ⲙⲙⲁⲉⲓ rest. T | ⲉϥϫⲓ ⲉⲛ] lac. T, to be restored as ⲉϥϫⲓ ⲉⲛ 18 ⲙⲡⲉⲧⲛⲛⲁ-] i.e. ⲙⲡⲉⲧⲛⲁ-, probably thus to be restored in T | ⲙⲁϥ] i.e. ⲙⲙⲁϥ T 19 ⲉⲧⲛⲛⲁ-] i.e. ⲉⲧⲛⲁ-; ⲡⲉⲧⲛⲁ- rest. T | ⲙⲁϥ] i.e. ⲙⲙⲁϥ rest. T 20 ⲉⲛ ⲡⲉⲛⲧⲁⲉⲓⲥⲉϫⲉ] ⲛⲧⲁⲉⲓⲥⲉϫⲉ ⲉⲛ T | ⲟⲩⲁⲉⲧ rest.] ⲟⲩⲁⲉⲉⲧ T 21 ⲧⲉⲩⲟⲩⲁⲉⲓ] ⲧⲉⲩⲁⲉⲓ T | ⲛⲏⲓ rest.] ⲛⲏⲉⲓ T 22 ⲡⲉⲧⲛⲛⲁ-] em. ⲡⲉϯⲛⲁ- T 23 ⲟⲩⲱⲱϩ rest.] i.e. ⲟⲩⲱⲱⲛϩ T 24 ⲛϣⲁ-] ϣⲁ- T | ⲙⲁⲩ] i.e. ⲙⲙⲁⲩ T 25 ⲛⲏⲓ] ⲛⲏⲉⲓ T 26 ϩⲁⲑⲏ+ⲇⲉ T | ⲛⲧ-] ⲙⲡ- T

p. 14* John 13:2-12

ⲟⲩ[ⲇ]ⲉ̣ⲓⲡⲛⲟⲛ̣ [ϣⲱⲡⲉ ⲉⲁⲡⲇⲓⲁⲃⲟⲗⲟⲥ ⲟⲩⲱ ⲉϥⲛⲟⲩ]
ϫⲉ̣ [ⲙ]ⲙⲁⲥ ⲁⲡ̣[ϩⲏⲧ ⲛⲓⲟⲩⲇⲁⲥ ⲡϣⲏⲣⲉ ⲛⲥⲓⲙⲱⲛ]
ⲡⲓⲥⲕⲁⲣⲓⲱⲧⲏ[ⲥ ϫⲉ ⲉϥⲁⲣⲡⲁⲣⲁⲇⲓⲇⲟⲩ ⲙⲁϥ ⲉϥⲥⲁⲩ] 13:3
ⲛⲉ ⲛ̅ϭⲓⲓ̅ⲥ̅ ϫⲉ [ⲁⲡⲓⲱⲧ ϯⲛⲕⲉⲉⲛ ⲛⲓⲙ ⲁϩⲣⲏⲓ ⲁ]
ⲛⲉϥϭⲓϫ ⲁⲩⲱ [ϫⲉ ⲛⲧⲁϥⲉⲓ ⲁⲃⲁⲗ ϩⲓⲧⲛⲡⲛⲟⲩⲧⲉ]
ⲁⲩⲱ ⲉϥⲛⲁ ⲁ̣ⲣ̣[ⲉⲧϥ ⲙⲡⲛⲟⲩⲧⲉ ⲁ]ϥ̣[ⲧⲱⲱⲛ ϩⲛ] 13:4
ⲡⲇⲉⲓⲡⲛⲟⲛ ⲁ̣ϥ̣[ⲕⲁⲛⲉϥϩⲁⲉⲓⲧⲉ ⲁϩ]ⲣ̣ⲏⲓ ⲁϥϫⲓ̣ⲟ̣ⲩ
ⲗⲉⲛⲧⲓⲟⲛ ⲁϥⲙ̣[ⲁⲣⲉϥ ⲙⲁϥ ⲁⲩⲱ ⲁϥ]ⲛ̣ⲁϫⲙⲁⲩ ⲁⲩⲧⲗⲗ 13:5
ⲕⲁⲛⲏ ⲁϥⲁⲣⲭ[ⲉⲓ ⲛⲓⲟⲩⲉ ⲛⲛⲟⲩⲣⲓⲧ]ⲉ ⲛ̅ⲛⲉϥⲙⲁⲑⲏ
ⲧ̣ⲏ̣ⲥ̣ [ⲁⲩⲱ ⲛϥⲁⲧⲟⲩ ⲙⲡⲗⲉⲛⲧⲓⲟⲛ ⲉⲧ]ⲙⲏⲣ ⲙⲁϥ ⲁϥⲓ ϭⲉ 13:6
[ϣⲁⲥⲓⲙⲱⲛ ⲡⲉⲧⲣⲟⲥ ⲡⲁϫⲉⲡⲉⲧⲙⲙ]ⲉ̣ⲩ ⲛⲉ[ϥ] ϫ̣ⲉ [ⲡ]
[ϫⲁⲉⲓⲥ ⲛⲧⲁⲕ ⲡⲉⲧⲛ]ⲁⲉⲓⲁⲛ̣ⲛ̣ⲁ̣[ⲟⲩⲣⲓⲧⲉ ⲁⲓ̅ⲥ̅ ⲟⲩⲱϣⲃ] 13:7
[ⲡⲁϫⲉϥ ⲛ]ⲉ̣ϥ ϫ̣[ⲉ] ⲡⲉ̣ϯⲉⲓⲣⲉ̣ ⲙⲁϥ ⲕⲥⲁⲩⲛⲉ
[ⲙⲁϥ ⲛⲧⲁⲕ ⲉⲛ ϯ]ⲛⲟⲩ ⲙⲛⲛⲥⲱⲥ ⲇⲉ ⲕⲛⲁⲙⲙⲉ
[ⲁⲣⲁϥ ⲡⲁϫⲉⲡ]ⲉⲧⲣⲟⲥ ⲛⲉϥ ϫⲉ ⲁⲛⲁⲕⲁⲁⲕ ⲁⲉⲓⲁ 13:8
[ⲣⲉⲧ ⲁⲛⲏϩⲉ ⲁⲓ]ⲥ̣̅ ⲟⲩⲱϣⲃ ⲡⲁϫⲉϥ ⲛⲉϥ ϫⲉ ⲉⲓⲧⲙⲉⲓ
ⲁ̣ⲣⲉⲧ[ⲕ ⲙⲛⲧⲉⲕⲙ]ⲉⲣⲟⲥ ⲛ̅ⲙⲏⲓ ⲡⲁϫⲉϥ ⲛ̣[ⲉ]ϥ ⲛ̅ 13:9
ϭⲓⲥⲓ[ⲙⲱⲛ ⲡⲉⲧ]ⲣⲟⲥ ϫⲉ ⲡϫⲁⲉⲓⲥ ⲟⲩⲙⲟⲛⲟⲛ
ⲛ̣ⲁ̣[ⲟ]ⲩ[ⲣⲓⲧⲉ ⲁⲗⲗ]ⲁ ⲛⲁⲕⲉϭⲓϫ ⲙⲛⲧⲁⲁⲡⲉ ⲡⲁ 13:10
ϫⲉⲓ̅ⲥ̅ ⲛⲉ̣[ϥ ϫ]ⲉ̣ ⲡⲉⲛⲧⲁϥϫⲱⲕⲙ ⲛϥⲣⲭⲣⲓⲁ
ⲉ̣ⲛ ⲉⲓⲙⲏ[ⲧⲓ ⲛϥ]ⲓⲁⲛⲛⲉϥⲟⲩⲣⲓⲧⲉ ⲁⲗⲗⲁ `ϥ´ⲧⲟⲩⲃⲁ

13:5 8 What has been transcribed as ⲩ in ⲛⲁϫⲙⲁⲩ gives an impression similar to the ⲩ corrected from ϥ in Jn 11:57 (p. 10:5, see note above). Obviously the scribe first wrote ⲛⲁϫⲙⲁϥ and then imposed the shape of an ⲩ (with two diagonal strokes) on the ϥ without erasing the upper end of the long hasta of ϥ, which is still visible.

13:10 21 Towards the end of the line, the scribe first wrote ⲁⲗⲗⲁⲧⲟⲩⲃⲁ and then inserted ϥ half above the line, with the lower end of its hasta crossing the link between ⲁ and ⲧ.

---

1 ⲇⲉⲓⲡⲛⲟⲛ] ⲇⲓⲡⲛⲟⲛ T 3 ⲙⲁϥ rest.] i.e. ⲙⲙⲁϥ T 6 ⲛⲁ] ⲛⲛⲁ T (partly rest.) 7 ⲇⲉⲓⲡⲛⲟⲛ] ⲇⲓⲡⲛⲟⲛ rest. T | ⲁϥⲕⲁⲛⲉϥ- rest.] probably thus to be restored in T 8 ⲙⲁϥ rest.] i.e. ⲙⲙⲁϥ rest. T 8/9 ⲁⲩⲧⲗⲗⲕⲁⲛⲏ] em. ⲁⲧⲗⲁⲕⲁⲛⲏ (rest. T) or ⲁⲩⲗⲁⲕⲁⲛⲏ 9 ⲛⲛⲉϥ-] ⲛⲙ- rest. T (probably not enough space for ⲛⲛⲉϥ-) 10 ⲛϥⲁⲧⲟⲩ rest.] thus to be restored in T | ⲙⲁϥ] i.e. ⲙⲙⲁϥ T 12 ⲡⲉⲧⲛⲁ- rest.] ⲡⲉⲧⲁ- T | ⲛⲛⲁ-] i.e. ⲛⲁ- T 13 ⲙⲁϥ] i.e. ⲙⲙⲁϥ T 14 ⲙⲁϥ rest.] i.e. ⲙⲙⲁϥ T 15 ⲁⲛⲁ-] probably em. ⲛⲁ- T (partly rest.) 16 ⲉⲓ-] ⲉⲉⲓ- T 17 ⲛⲙⲏⲓ] i.e. ⲛⲙⲙⲏⲓ; ⲛⲙⲙⲏⲉⲓ T 21 ⲛⲛⲉϥ-] i.e. ⲛⲉϥ- T

ⲉⲓⲧ ⲧⲏⲣϥ̄ [ⲛⲧ]ⲱ̣ⲧⲛ ϩⲱⲟⲩⲧⲛ ⲧⲉⲧⲛⲧⲟⲩⲃⲁⲉⲓⲧ
ⲁⲗⲗⲁ ⲧ̣ⲱ̣[ⲧⲛ ⲉⲛ] ⲧⲏⲣⲧⲛ ⲛⲉϥⲥⲁⲩⲛⲉ ⲅⲁⲣ ⲙ̄ⲡⲉⲧⲛ 13:11
ⲛⲁⲣ̣ⲡⲁⲣ[ⲁⲇⲓⲇⲟⲩ] ⲙⲁϥ ⲉⲧⲃⲉⲡⲉⲉⲓ ⲁϥϫⲟⲟⲥ
[ϫⲉ ⲧⲉⲧⲛⲧⲟⲩ]ⲃ̣ⲁⲉⲓⲧ̇ ⲧⲏⲣⲧⲛ ⲉⲛ ⲛ̄ⲧⲁⲣⲉϥⲟⲩ 13:12
[ⲱ ⲇⲉ ⲉϥⲉⲓⲟⲩⲉ ⲛⲛⲟⲩⲟⲩⲣ]ⲓ̣ⲧⲉ ⲁϥϫⲓⲛⲉϥ
[ϩⲁⲉⲓⲧⲉ ⲁϥⲛⲁϫϥ ⲁⲛ ⲁⲩ]ⲱ̣ ⲡⲁϫⲉϥ ⲛⲉⲩ
[ϫⲉ ⲧⲉⲧⲛⲥⲁⲩⲛⲉ ϫⲉ ⲉⲩ ⲡⲉⲛ]ⲧ̣ⲁⲉⲓⲉϥ ⲛ̣[ⲏ]

13:10 23 The letters following ⲁⲗⲗⲁ can be identified as either ⲧⲱ or ⲡ plus another letter. In the present context, therefore, one can safely assume that the scribe wrote ⲧⲱⲧⲛ (for ⲛⲧⲱⲧⲛ; cf. ⲧⲁⲕ, ⲧⲁϥ for ⲛ̄ⲧⲁⲕ, ⲛ̄ⲧⲁϥ), with the second ⲧ being no longer visible on the papyrus: on the right-hand margin of this fragment, there must have been letters which are no longer visible also in the lines above and below this line. The lacuna, then, provides sufficient space for ⲉⲛ, which is not found where T has it (after ⲧⲏⲣⲧⲛ).

13:12 26 The spelling ⲉϥⲉⲓⲟⲩⲉ has been restored from T. The irregularity of the hand hardly permits one to determine the precise number of letters in a lacuna of this size. The form used by the scribe may well have been ⲉϥⲓⲟⲩⲉ, in accordance with the general principles of orthography in this MS (see Introduction).

---

23 ⲧⲱⲧⲛ ⲉⲛ ⲧⲏⲣⲧⲛ] i.e. ⲛⲧⲱⲧⲛ etc.; ⲧⲏⲣⲧⲛ ⲉⲛ T 23/24 ⲡⲉⲧⲛⲛⲁ-] i.e. ⲡⲉⲧⲛⲁ- T
24 ⲙⲁϥ] ⲙⲙⲁϥ rest. T 25 ⲧⲏⲣⲧⲛ ⲉⲛ] ⲉⲛ ⲧⲏⲣⲧⲛ T 28 ⲉϥ] ⲉⲉϥ rest. T

p. 15* John 13:12-27

[τν   ντωτν τετνμογτε α]ρ̣[αε]ι̣ ϫε̣ [πcαϩ αγω] 13:13
[πϫαειc αγω καλωc τετ]νϫω μμα̣[c ανακ γαρ πε]
[εϣωπε ανακ αει(ει)ανετ]νογριτε π̣[ϫαειc ...] 13:14
[ ]....[ ]
[missing 13:16
[
[εϣωπε] τ̣[ε]τ̣ν̣[cαγνε ννεει] 13:17
[νεειετντηνε ετετνϣ]α̣ετογ εε̣ι̣ϫερ̣[ωτν] 13:18
[τηρτν εν ανακ ετcαγν]ε ννενταειc[ατπογ]
[αλλα ϫεκαcε ερετγραφ]η̣ ναϫωκ̣ α̣β̣[αλ ...]

(lines 11 to 16 missing)

13:22
[ν]ϭ̣ι̣μμ̣[αθητηc εγρμαει]
[ϩε ϫε εϥ]ϫ̣ω μαc ετβενιμ ν̣[εγνογεε αε] 13:23
[αβαλ ϩ]ν̣νεϥμ̣αθητηc εϥν̣[ηϫ ϩνκογνϥ]

13:13-38 pp. 15*/16* The amount of text missing between the remainders of the fourth line (fragment a) on each of these pages (illegible on p. 15) and the traces of the uppermost readable lines on fragment b suggest that the number of lines that are completely missing should be one higher for p. 16 than for p. 15. Although it may appear strange that the same fragment b should contain, as indicated in the transcription, lines 7-10 on the recto, but lines 8-11 on the verso, this assumption seems to be justified by the fact that line 15:7 is about half a line higher on the fragment than 16:8, and 15:7-10 have more space between lines than do 16:8-11. Furthermore, the fact that the first three lines of fragment c (i.e., lines 17-19) are about half a line higher on p. 16 than they are on p. 15 may be due to there being one more line of text between fragments b and c on p. 15 than on p. 16. — The text restored in the lacunae is largely that of T. As the vertical alignment of the three fragments is merely hypothetical, the assortment of text to lines is only exempli gratia.

13:23 19 The shorter spelling κογνϥ, instead of κογογνϥ (T), is suggested by the space available between the remainders of lines 19 and 20 (irrespective of the actual line division). The rest of the restoration is based on the assumption that the same text can be restored in T (left empty by Thompson): p. 68, lines 16-17 νι[ηc πε] | [τε ιηc μα]ειε.

---

3 εϣωπε rest.] εϣπε T   7 εϣωπε rest.] εϣπε T   8 ετετνϣα- rest.] ερεϣατετν- T | ετογ] εετογ T   19 ϩν-² rest.] same rest. T, but α- likewise possible | κογνϥ rest.] κογογνϥ T

[ⲛⲓ̅ⲥ̅ ⲡⲉⲧⲉⲓ̅ⲥ̅] ⲙⲁⲉⲓⲉ ⲙⲁϥ   ⲁ[ϥϫⲱⲣⲙ ϭⲉ ⲟⲩ] 13:24
[ⲃⲉⲡⲉⲉⲓ ⲛ]ϭⲓⲥⲓⲙⲱⲛ ⲡⲉ̣ⲧ[ⲣⲟⲥ ⲁϫⲛⲟⲩϥ]
[ϫⲉ ⲛⲓⲙ ⲡ]ⲉⲧⲉϥⲥⲉϫⲉ ⲁⲣ̣[ⲁϥ   ⲁ] 13:25
[ⲡⲏ ϭⲉ ⲛⲁ]ϫ̣ϥ ⲁϫⲛⲧⲙⲉⲥ̣ⲧ̣ϩⲏⲧ ⲛ̣[ⲓ̅ⲥ̅ ⲡⲁϫⲉϥ]
[ⲛⲉϥ ϫⲉ ⲡ]ϫ̣ⲁ̣[ⲉⲓⲥ ⲛⲓ]ⲙ ⲡⲉ   ⲁϥ[ⲟⲩⲱϣⲃ ⲛϭⲓ] 13:26
[ⲓ̅ⲥ̅ ϫⲉ ⲡⲉϯⲛⲁⲥⲁⲡ]ⲡ̣ⲁⲉⲓⲕ ⲛ̄[ⲧⲁⲧⲉϥ ⲛⲉϥ ⲡⲉ]
[ⲧⲙⲙⲉⲩ ⲡⲉ ⲁⲩⲱ ⲁϥⲥⲁⲡ]ⲡⲁ̣ⲉ̣ⲓ̣[ⲕ ⲁϥⲧⲉϥ ⲛ]
[} missing
[} 13:27

20 ⲙⲁϥ] i.e. ⲙⲙⲁϥ T (partly rest.)   22 ⲡⲉⲧⲉϥ-] ⲡⲉⲧϥ- rest. T   25 ⲧⲉϥ rest.] ⲧⲉⲉϥ rest. T

p. 16* John 13:27-38

[ⲃⲱⲕ ⲁϩⲟⲩⲛ ⲁ]ⲣ̣[ⲁ]ϥ [ⲡⲁϫⲉϥ ϭⲉ ⲛⲉϥ ⲛϭⲓ ⲓ̅ⲥ̅ ϫⲉ ⲡⲉⲧⲕ]
[ⲛⲁⲉϥ ⲁⲣ]ⲓⲣⲉ ϩⲛⲛⲟⲩ[ϭⲗⲁⲙ ⲙⲡⲉⲗⲁⲩⲉ ⲇⲉ ⲛⲛⲉⲧⲛⲏϫ] 13:28
[ⲛⲙⲙⲉϥ] ⲙⲙⲉ ϫⲉ ⲉ[ⲧⲃⲉⲉⲩ ⲁϥϫⲉⲡⲉⲉⲓ ⲛⲉϥ ⲛⲉⲣⲉϩⲁⲉⲓⲛⲉ] 13:29
[ⲇⲉ ⲙⲉⲩⲉ] ϫⲉ [ⲉⲡⲓⲇⲏ ...
[
[} missing
[ⲛⲧⲁⲣⲉϥⲉⲓ ⲁ] 13:31
[ⲃⲁⲗ ⲡⲁ]ϫ̣[ⲉ]ϥ̣ [ⲛϭ]ⲓ̣ⲓ̣ⲥ̣ ϫⲉ ϯ[ⲛⲟⲩ ⲁϥϫⲓⲉⲁⲩ ⲛϭⲓ ⲡϣⲏⲣⲉ ⲡ]
[ⲣⲱⲙⲉ] ⲁⲩⲱ ⲁ̣ⲡⲛⲟⲩⲧⲉ [ϫⲓⲉⲁⲩ ϩⲣⲏⲓ ⲛϩⲏⲧϥ ⲁⲩⲱ ⲡⲛⲟⲩⲧⲉ] 13:32
[ⲛⲁ]ϯⲉⲁⲩ ⲛⲉϥ ⲛϩⲏⲧϥ [ⲁⲩⲱ ⲛⲧⲟⲩⲛⲟⲩ ϥⲛⲁϯⲉⲁⲩ ⲛⲉϥ ⲛⲁ] 13:33
[ϣⲏⲣ]ⲉ ⲉⲧ̣ⲓ̣ ⲕ̣ⲁ̣ⲓⲕⲟⲩⲉⲓ ⲛ̣[ⲟⲩⲁⲉⲓϣ ...

(lines 12 to 16 missing)

[ϩⲛⲡⲉ]ⲉ̣ⲓ̣ [ⲟⲩⲁⲛ ⲛⲓⲙ ⲛⲁⲙⲙⲉ ϫⲉ] 13:35
[ⲛⲧⲱⲧⲛⲛ]ⲁ̣ⲙⲁⲑⲏⲧ̣ⲏⲥ ⲉⲣⲉϣⲁ̣ⲙ̣ⲣ̣[ⲡⲉⲛⲉⲧⲛⲉⲣⲏⲩ]
[ⲡⲁϫⲉϥ ⲛⲉϥ] ⲛϭⲓⲥⲓⲙⲱⲛ ⲡⲉⲧⲣⲟ̣ⲥ̣ ϫ̣[ⲉ ⲡϫⲁⲉⲓⲥ ⲉⲕⲛⲛⲁ] 13:36
[ⲁⲧⲟ ⲁϥⲟⲩⲱ]ϣ̣ⲃ ⲛⲉϥ ⲛϭⲓ ⲓ̅ⲥ̅ ϫⲉ ⲡ̣[ⲙⲁ ⲁⲛⲁⲕ ⲉϯⲛⲛⲁ]
[ⲁⲣⲁϥ ⲕⲛⲁϣⲟⲩⲁ]ϩ̣ⲕ ⲛ̄ⲥⲱⲉⲓ ⲉⲛ ⲡⲁ̣[ϫⲉϥ ⲛⲉϥ ϫⲉ ⲡϫⲁ] 13:37
[ⲉⲓⲥ ⲉⲧⲃⲉⲉⲩ ϯⲛ]ⲁϣⲟⲩⲁ`ϩⲧ´ ⲉⲛ ⲥⲱⲕ ϯ[ⲛⲁⲕⲟⲩ ϯⲛⲟⲩ]
[ⲛⲧⲁⲯⲩⲭⲏ ⲁ]ϩ̣ⲣ̣ⲏ̣ⲓ ϩⲁⲣⲁⲕ ⲁϥⲟⲩⲱϣⲃ̣ [ⲛϭⲓ ⲓ̅ⲥ̅ ϫⲉ ⲕⲛⲁⲕⲁ] 13:38
[ⲧⲉⲕⲯⲩⲭ]ⲏ ϩⲁ̣ⲣⲁⲉⲓ ϩⲁ[ⲙ]ⲏ̣ⲛ ϩⲁⲙ[ⲏⲛ ϯϫⲱ ⲙⲙⲁⲥ]
[ⲛⲉⲕ ϫⲉ] ⲛ̣ⲉⲟⲩⲁⲗ̣[ⲉⲕⲧⲱⲣ ⲙⲟⲩⲧⲉ ⲉⲙⲡⲁⲧⲕⲣⲁⲣ]
[ⲛⲁ ⲙ]ⲙⲁⲉⲓ [ⲛϣⲁⲙⲧ ⲛⲥⲁⲡ ...

(probably two lines missing)

13:37 22 The letters actually found in the line are ⲁϣⲟⲩⲁⲕⲉⲛ etc. The two letters ϩⲧ are written above the ⲁ in ⲟⲩⲁⲕ. The letter ⲕ gives the impression of being slightly erased, although this is not clear.

---

2 ⲉϥ rest.] ⲉⲉϥ T | ⲁⲣⲓⲣⲉ rest.] ⲉⲣⲓϥ T | ϩⲛⲛ-] i.e. ϩⲛ- T 4 ⲙⲉⲩⲉ rest.] ⲙⲉⲉⲩⲉ T 8 ⲡ- rest.] em. ⲙⲡ- T 10 ⲛⲉϥ+ϩⲣⲏï T 11 ⲕⲁⲓⲕⲟⲩⲉⲓ ⲛⲟⲩⲁⲉⲓϣ partly rest.] ⲕⲁⲓⲟⲩⲁⲉⲓϣ ϣⲏⲙ T 18 ⲉⲣⲉϣⲁ-] em. ⲉⲣⲉϣⲁⲧⲉⲧⲛ- or ⲉⲣⲉⲧⲉⲧⲛϣⲁ-; ⲉⲧⲉⲧⲛϣⲁ- T 22 ϯ- rest.] ⲛϯ- T | ⲥⲱⲕ] i.e. ⲛⲥⲱⲕ T

INDICES

(An asterisk "*" indicates that the word is partially restored).

COPTIC WORDS

(ⲟⲩⲱϣ), ^{1}ⲟⲩϣ, ^{2}ⲟⲩⲱϣ- vb tr. *love, want* 11:^{2}5; 12:^{1}21
ⲟⲩϣⲓ nn f. *night* 11:10
ⲟⲩⲱϣⲃ, ^{1}ⲟⲩϣⲃ vb tr. *answer* 10:32, 33, 34; 11:9; 12:^{1}23*, ^{1}34; 13:8, 36*, 38
ⲟⲩⲱϣⲧ vb intr. *worship* 12:20
(ⲟⲩⲱϩ), ⲟⲩϩ vb tr. *place*
  ⲟⲩϩ ⲥⲱ⸗ *follow* 12:26
  ⲟⲩⲁϩ⸗ (refl.) (ⲛ)ⲥⲱ⸗
    ⲟⲩⲁϩⲧ 13:37
    ⲟⲩⲁϩⲕ 13:36*
    ⲟⲩⲁϩⲟⲩ 10:27; 11:31
ⲟⲩϫⲉⲉⲓ vb intr. *be saved* 10:9

ⲱⲃⲉϣ nn m. *forgetfulness, sleep* 11:13
ⲱⲛⲉ nn m. *stone* 10:31, 32, 33; 11:8, 38, 39, 41
(ⲱϣ) vb *announce*
  ⲁϣ-, ϣ-, see ϭⲏⲗ
ⲱϣ nn m. *thing* 10:33
ⲱⲱϩ, ^{1}ⲱϩ, ⲁⲁϩ† vb intr. *live* 11:25, 26†
  nn m. 10:10, 28; 11:^{1}25; 12:50*
(ⲱϩⲉ), ⲁϩⲉ† vb intr. *stand*
  ⲁϩⲉ ⲁⲣⲉⲧ⸗
    ⲁϩⲉ ⲁⲣⲉⲧϥ 11:42; 12:29*
    ⲁϩⲉ ⲁⲣⲉⲧⲟⲩ 11:56*

ϣ- vb *be able to* 10:21; 13:37
ϣⲁ- preposition *to* 10:28; 11:19, 26, 45, 46; 12:32, 34, 50
  ϣⲁⲣⲁ⸗
    ϣⲁⲣⲁⲉⲓ 12:32*
    ϣⲁⲁⲣⲁϥ 10:41; 11:29
ϣⲁⲣⲁϥ 11:3, 15
ϣⲁⲣⲁⲏ 10:35*
ϣⲁⲉⲓⲉ nn m. *festival* 11:56; 12:12, 20, 20; 13:1
ϣⲱⲉ nn *hundred*, see ϣⲁⲙⲧ
ϣⲃⲏⲣ nn m. *friend* 11:11
  ϣⲃⲣ- 11:16*
ϣⲏⲙ adj. *small* 12:35
(ϣⲁⲙⲧ), ϣⲙⲧ- nn *three*
  ϣⲙⲧϣⲱⲉ *three hundred* 12:5*
ϣⲓⲛⲉ vb tr. *seek*
  ϣⲓⲛⲉ (ⲛ)ⲥⲁ- with infinitive *seek to* 10:39; 11:8, 56
ϣⲱⲛⲉ vb intr. *be sick* 11:1, 2, 3, 6
  nn m. 11:4
ϣⲱⲡ vb tr. *receive*
  ϣⲡ-, see ϩⲙⲁⲧ
ϣⲱⲡⲉ, ϣⲟⲡ† vb intr. *become* 10:16, 19, 22, 35, 38†; 11:4†, 10†, 54; 12:26, 29*, 34, 35†, 36, 42
ϣⲉⲉⲣⲉ nn f. *daughter* 12:15

ϣⲏⲣⲉ nn m. *son, child* 10:36*; 11:4*, 27, 52; 12:23, 34*, 36*; 13:33*
ϣⲁⲣⲉⲡ nn
  ⲛϣⲁⲣⲉⲡ adverb *at first* 10:40; 12:16*
ϣⲱⲥ, ^{1}ϣⲥ nn m. *shepherd* 10:^{1}11, ^{1}11, 12, ^{1}14, 16
ϣⲟⲩⲟⲩⲧ vb tr. *kill* 10:10
ϣⲧⲁⲣⲧⲣ vb intr. *be troubled* 11:33; 12:27

ϥⲓ vb tr. *take, carry* 10:18, 24; 11:48
  ϥⲓ- 10:31; 11:41
  see also ⲓⲉⲧ⸗
ϥⲟⲩⲉ nn m. *hair* 11:2; unclear 12:3
ϥⲱⲧⲉ, ^{1}ϥⲛⲧ- vb tr. *wipe* 11:^{1}2*; 12:3*
ϥⲧⲁⲩ nn *four* 11:39

ϩⲁ- preposition *below*, etc. 10:11, 13, 15, 23, 38; 11:32, 50, 51, 55; 12:1; 13:1
    ⲁⲃⲁⲗ ϩⲁ- 12:5
  ϩⲁⲣⲁ⸗
    ϩⲁⲣⲁⲉⲓ 10:18; 12:49*; 13:38
    ϩⲁⲣⲁϥ 11:51
    ϩⲁⲣⲁⲥ 12:7
    ⲁϩⲣⲏⲓ ϩⲁⲣⲁⲕ 13:37
    ⲁϩⲣⲏⲓ ϩⲁⲣⲁⲥ 12:35
ϩⲁⲉ nn m. *end*
  as adj. *last* 11:24
ϩⲁⲉⲓⲉ vb intr. *fall* 12:24*
  ϩⲁⲉⲓⲉ ⲁⲃⲁⲗ *perish* 10:28; 11:50
ϩⲉ nn f. *way* 11:36
  ⲕⲁⲧⲁⲑⲉ 10:15; 12:14, 50
  ⲛⲉϣ ϩⲉ 12:34
  ⲛⲑⲉ 11:33
  ⲛⲧⲉⲉⲓϩⲉ 11:48
ϩⲏ nn f. *front*
  ϩⲏⲧ⸗
    ϩⲏⲧϥ 11:20; 12:18*
  ϩⲁⲑⲏ *before* 11:55; 12:1; 13:1
ϩⲓ- preposition *at, on, in* 11:38
  see also ⲣⲉⲧ⸗, ⲡⲁϩⲟⲩ, ϩⲓⲧⲛ-
ϩⲟ nn m. *face* 11:44
(ϩⲟⲩⲟ), ϩⲟ nn m. *greater part* 10:10
ϩ(ⲱ)ⲟⲩ⸗ pronoun reinforcer
  ϩⲟⲩⲧ 12:32
  ϩⲱⲟⲩⲛ 11:16
  ϩⲱⲟⲩⲧⲛ 13:10
ϩⲱⲃ nn m. *thing, deed* 10:32, 32; 11:22
  ϩⲃⲏⲩⲉ, ^{1}ⲃⲏⲩⲉ pl. 10:25, ^{1}37, ^{1}38
ϩⲏⲕⲉ nn m. *poor* 12:5, 8
ϩⲙⲁⲥⲧ, ^{1}ϩⲛⲙⲁⲥ vb intr. *sit* 11:^{1}20; 12:15*
ϩⲙⲁⲧ nn m. *grace*
  ϣⲡϩⲙⲁⲧ *thank* 11:41
ϩⲛ-, ^{1}ϩⲛⲛ- preposition 10:19, 22, 22, 23, 25, 34, 39*; 11:9, 10, 17, 20, 21, 24, 30,

GREEK WORDS

PROPER NOUNS

CONJUGATIONS

Bipartite

I Present

Circumstantial

Imperfect

Relative I Present

II Present

I Future

Future Imperfect

Relative I Future

II Future

Tripartite

I Perfect

ORIENTALISTE, P.B. 41, B-3000 Leuven